The Physical Basis of Biochemistry

Peter R. Bergethon · Kevin Hallock

The Physical Basis
of Biochemistry

Solutions Manual to the Second Edition

 Springer

Peter R. Bergethon
Departments of Anatomy & Neurobiology
 and Biochemistry
Boston University School of Medicine
Boston, MA 02118-2526, USA
prberget@bu.edu

Kevin Hallock
Department of Anatomy
 and Neurobiology
Boston University School of Medicine
Boston, MA 02118-2526, USA
hallockk@bu.edu

ISBN 978-1-4419-7363-4   e-ISBN 978-1-4419-7364-1
DOI 10.1007/978-1-4419-7364-1
Springer New York Dordrecht Heidelberg London

Library of Congress Control Number: 2010937181

Printed on acid-free paper

Springer is part of Springer Science+Business Media (www.springer.com)

Preface

Physical studies are really only learned by doing and struggling with problems. Every professor knows this and every student fears it. Problems are hard enough in courses where the main goal is to ensure familiarity with the major tools used in the discipline. In biophysical chemistry the problems are somewhat more difficult because not only is the student struggling with formulas and concepts but the questions and problems are often nuanced, deeply nested and complex.

We wrote this small manual as a companion to the new Edition of *The Physical Basis of Biochemistry: Foundations of Molecular Biophysics*. Our intention is to provide the students who are taking the course experience with solving problems and thinking about the concepts in the course without being overwhelmed. A fair number of the problems are straightforward but these are balanced with some that are real world and challenging. We know from using problems in our own teaching that a few questions that force thinking and analysis rather than only rote "drill and kill" lists are best for teaching the topics covered in biophysical chemistry.

Not every topic in the main textbook is covered in the solutions manual and we have not made this manual exhaustive in terms of complete coverage or overwhelming numbers of questions on every chapter. Instead we have tried to be judicious in choosing topics and scenarios that support teaching and learning and that will often take time and thought to accomplish. We hope that we have struck the balance that will encourage students to do the several problems and appreciate the depth that most answers explore rather than see the manual as an exercise obligation.

We did recognize as we worked through each problem ourselves that it is sometimes easy to expect one response to a question but instead to serve only confusion to the person solving it. We have tried to capture all errors, both computational and those generating confusion. It is unlikely that we have done so and we encourage all users to inform us of errors, confusion and also to look for other materials that will support this Solutions Manual and the broader course.

Three words of advice:

- Do, do the problems. This alone will help you learn the material and use it in your research and scientific life.
- Pay attention to dimensional analysis. This is the trick to understanding and to checking your own developing expertise. If you do nothing else, do the dimensional analysis on these problems. Biophysical studies are hard because you can get lost. Dimensional analysis is the map. Use it.
- Have fun. Really. We did when we wrote and solved these problems. And stick with it. It is worth the trouble to become more expert.

Boston, Massachusetts   Peter R. Bergethon
June 2010   Kevin Hallock

Contents

Part I
Principles of Biophysical Inquiry

Chapter 1
Philosophy and Practice of Biophysical Study

1.1 Questions

Q.1.1 A system can be described by listing its system components, the (1) overall or "emergent" properties, (2) elements that comprise it, (3) the way the elements are related to one another and to the background or context space, (4) the characteristics of the contextual space. A *graphical organizer* can be very useful to summarize these system components.

Design a graphical organizer(s) for the description of a system or structure and its properties.

Q.1.2 Use a graphical organizer to describe the following system:

Q.1.3 Write a systems description for a familiar scenario such as a sports event or game.

Q.1.4 Common problems in scientific investigation are epistemological in nature. Where in the progression of inquiry are most epistemological problems located?

1.2 Thought Assignment

For each model system developed in this book, make it a habit to write out the systems description whenever you encounter that model. This includes the kinetic theory of gases, thermodynamic systems, the Born model, the Debye-Hückel model, electric circuit models of electrochemical systems, etc.

This chapter from *The Physical Basis of Biochemistry: Solutions Manual to the Second Edition* corresponds to Chapter 2 from *The Physical Basis of Biochemistry, Second Edition*

1.3 Answers

A.1.1 A useful graphical organizer that represents properties as emergent from the systemic structure which is comprised of elements, rules and boundary/background space.

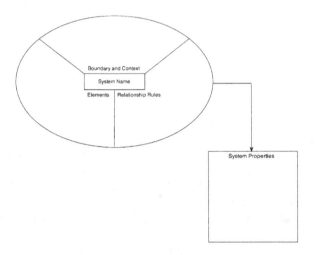

A.1.2 Start with an overall system analyzer that shows the properties of the pattern or structure.

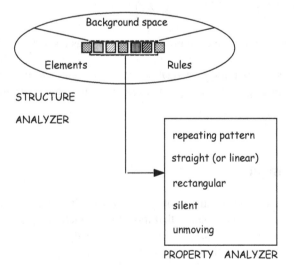

A more detailed analysis of the structure of the system parts can be written:

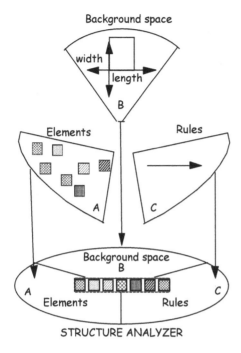

The elements can be further analyzed with an element analyzer.

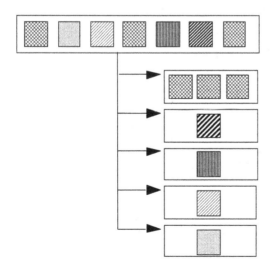

A.1.3   A sports event or game is a structured system that can be described as follows (the emergent properties of the overall system are not pictured here):

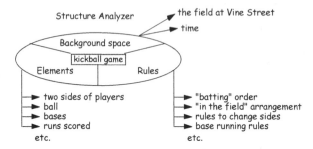

A detailed element analyzer of the kickball can be performed.

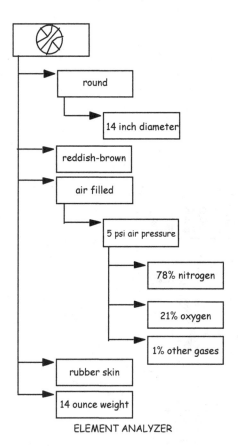

A.1.4 A prominent role for epistemological study comes in the experimental con-
       sideration (and skeptical evaluation of data) relating to the formation of the
       original descriptive model drawn from observation. Much excitement exists
       around "hypothesis-oriented" scientific investigation in which descriptive
       models are transformed into proposed linked hypotheses thus generating the
       explanatory or theoretical model. These theoretical models generate predic-
       tions that can be tested by well-designed experimental models that compare
       the experiment with the original description of reality. However the descrip-
       tive model itself should be carefully investigated to be sure that it is not
       artifact or a product of observer-reality coupling that leads to a poor descrip-
       tive model. It is in this careful empirical consideration that epistemological
       concerns enter scientific investigation.

Chapter 2
Overview of the Biological System
Under Study – Descriptive Models

2.1 Thought exercises

This chapter is intended largely as a survey of the biological system in which our physical studies occur. The following exercises may be useful for students who are unfamiliar with much of the basic content in biological studies to help review the chapter content. Because the answers are essentially a recap and reorganization of the chapter content itself, answers are not given in this manual.

Q.2.1 Use a graphical organizer (see Chapter 1) to write a systems analysis of

    (a) a prokaryotic cell,
    (b) an eukaryotic cell.

Q.2.2 Extend your systems analysis to various subsystems of the cell including (1) the cytosol, (2) the ribosome; (3) mitochondrion; (4) cell membrane; (5) nucleus.

Q.2.3 Are the sub-systems described in Question 2.2 the same for the prokaryotic cells and for the eukaryotic cells?

Q.2.4 Graphical organizers can be useful in describing changes that occur in complex systems. Use a "change" organizer to summarize the phenomenon of the "rusting of the earth" that occurred around 2 billion years ago.

Q.2.5 Consider the lipid membranes in the cell. Which organellar membranes are equivalent?

Q.2.6 Does the endosymbiotic theory support the view that compartmentalization is causally related to (a) the surface-volume problem or (b) the oxygenation-energy catastrophe?

This chapter from *The Physical Basis of Biochemistry: Solutions Manual to the Second Edition* corresponds to Chapter 3 from *The Physical Basis of Biochemistry, Second Edition*

P.R. Bergethon, K. Hallock, *The Physical Basis of Biochemistry*,
DOI 10.1007/978-1-4419-7364-1_2, © Springer Science+Business Media, LLC 2011

Chapter 3
Physical Thoughts, Biological Systems – The Application of Modeling Principles to Understanding Biological Systems

3.1 Questions

Q.3.1 Compare the pre-Copernican model of geocentricism and the subsequent model of heliocentricism in terms of the coupling assumptions between observer and the observed.

Q.3.2 List several observables that will provide information about the metabolic state of a liver cell.

Q.3.3 List three observables characterizing the state of a muscle cell in terms of its metabolic activity.

Q.3.4 In modern MRI (magnetic resonance imaging) and SPECT (single photon emission computerized tomography) scans, the instruments measure the amount of blood flowing to a specific area of the brain (SPECT) or the amount of oxygen extracted from blood (the BOLD signal in functional MRI) in order to assess the "intellectual" use of that part of the brain.

What are the abstractions of this process that allow conclusions to be drawn? Are there likely to be any bifurcations or surprises in the linkages in your proposed state space?

Q.3.5 The PET (positron emission tomography) scanner uses a tagged radioisotope to measure glucose delivery to similar regions of the brain during "brain tasks". Is this a better system for observing "intellectual" function than those described in Question 3.4?

Q.3.6 List three central points explaining why modeling is important to scientific investigation.

Q.3.7 What links the observer with reality?

Q.3.8 Why do we normally use approximate laws?

Q.3.9 What is a bifurcation point?

Q.3.10 What causes "complexity" in a system?

Q.3.11 List three types of attractors that describe dynamic behavior.

This chapter from *The Physical Basis of Biochemistry: Solutions Manual to the Second Edition* corresponds to Chapter 4 from *The Physical Basis of Biochemistry, Second Edition*

P.R. Bergethon, K. Hallock, *The Physical Basis of Biochemistry*,
DOI 10.1007/978-1-4419-7364-1_3, © Springer Science+Business Media, LLC 2011

3.2 Thought Exercise

The practice of medical diagnosis is concerned with how health and disease can be characterized in terms of the linkages between observables similar to those discussed in Questions 3.2 and 3.3. Arrange the observables and linkages into an equation of state that reflects the "health" state of the system.

3.3 Thought Exercise

A prominent biochemist has been quoted as arguing in curriculum discussions that: *a modern biochemist does not need to know any biophysical chemistry in order to be successful.* Without quibbling over the term successful, explain why such a statement may be regarded as true or false. Proceed with your analysis in terms of systems theory and explain your reasoning with precise explication of the state space you are discussing, its observables, linkages, errors and surprises.

[Hint: Whatever viewpoint you choose to argue, the argument will be more easily made if you make a specific biological reference and relate your formal argument to a real system]

3.4 Answers

A.3.1 The pre-Copernican models firmly placed human existence at the center of the universe and hence had a highly coupled relationship between humans and the behavior of the universe. The heliocentric model shifts the observer-observed coupling to a weaker relationship.

A.3.2 The liver is an organ with many functions and the very definition of its metabolic state can be very complicated. Some of the major functions of the liver are to produce enzymes to break down fats and proteins and the conversion of sugars into proteins and vice versa. The liver makes most amino acids and will process nitrogen waste to make urea, which is excreted in the urine. The liver detoxifies many materials that are ingested and will metabolize drugs and alcohol. It stores certain vitamins, breaks down hemoglobin (from red blood cells) and maintains the level of glucose in the blood. It also makes 80% of the cholesterol in the body.

Useful observables that are informative about the state of the liver include the level of glucose in the blood, the amount of the protein albumin in the blood (produced by the liver), the level of liver enzymes in the blood (these are normally found in the liver cells and only get into the blood in elevated levels when the liver cells are diseased and release the enzymes). The activity of the detoxifying enzymes (called cytochrome p450 enzymes) can also reflect the state of the liver. The quantity of triglycerides measured in the blood can reflect lipid metabolism by the liver cells.

A.3.3   The muscle cell uses glucose and oxygen as energy sources in order to provide mechanical action. The metabolic state is reflected in the input-output observables of this energy demand. Lactic acid, NAD^+/NADH ratio, oxygen tension, pH, glucose, myoglobin-oxygen binding saturation are all observables that are informative about the muscle cell metabolic state.

A.3.4   Key abstractions are that blood flow or changes in blood flow can be linked to an understanding of the processes of intellectual computation. While the abstraction that cells that are involved in neural computing are using energy (information is organized and to keep it meaningful requires energy just to fight the tendency for entropy to increase and render it meaningless), there are likely many linkages that are unknown and perhaps unknowable in the proposed causal chain. This is certainly true for blood flow as well as oxygen extraction from the blood. Still these studies are state of the art and provide much information. There are likely to be bifurcations and surprise as the details of the causal chains are learned.

A.3.5   Glucose and oxygen are the essential substrates for biological energy production. All of the concerns with the SPECT and MRI abstractions will hold with PET scan studies. There is some advantage to having different observables to measure "intellectual tasks" however the equation of state will invariably link blood flow, glucose and oxygen so these are variably coupled observables into the "intellectual state of mind".

A.3.6   Choose three from the following list:

What is there about the nature of natural systems that requires modeling?
What is the nature of the observable quality and quantity?
How can we prevent from having to build unnecessary models?
How can we recognize systems and models that are similar?
Can we generalize our models to find and describe "laws of nature"?

A.3.7   Observables.

A.3.8   Our knowledge is usually incomplete because we cannot measure the entire system accurately.

A.3.9   The transition point from one family of equivalent forms to another family of different forms.

A.3.10 A system's behavior is called "complex" when the overall system's behavior is not predicted or accounted for based on its subsystems, often caused by unrecognized bifurcations.

A.3.11 Static, periodic, and strange attractors.

Chapter 4
Probability and Statistics

4.1 Questions

Q.4.1 Would a base 4 or 5 system of numbering be beneficial in devising mathematical methods for handling problems dealing with information transfer in DNA and RNA research?

Q.4.2 What is the total number of configurations possible in a polymer chain composed of 150 monomers each of which can take one of four conformations?

Q.4.3 Show that $\binom{n}{r} = \binom{n}{n-r}$.

Q.4.4 Show that $\binom{n}{1} = n$.

Q.4.5 Show that $\binom{n}{n} = 1$.

Q.4.6 Graphically show the distribution function of a uniformly distributed random variable.

Q.4.7 You have 20 amino acids available. How many pentapeptides can be made if an amino acid is not replaced after being used?

Q.4.8 You have 20 adenosines, 50 thymidines, 15 guanosines, and 35 cytosines. Pick 15 sequential nucleosides. What is the probability of getting 5 thymidine, 5 adenosine, 5 cystosine and 1 guanosine?

Q.4.9 Evaluate: $\ln 1.8 \times 10^{18}!$

Q.4.10 Prove that the binomial distribution and the Poisson distribution are nearly equivalent.

$$\binom{n}{k} p^k (1-p)^{n-k} = \frac{n(n-1)\cdots(n-k+1)}{k!} p^k (1-p)^{n-k}$$

This chapter from *The Physical Basis of Biochemistry: Solutions Manual to the Second Edition* corresponds to Chapter 5 from *The Physical Basis of Biochemistry, Second Edition*

P.R. Bergethon, K. Hallock, *The Physical Basis of Biochemistry*,
DOI 10.1007/978-1-4419-7364-1_4, © Springer Science+Business Media, LLC 2011

Q.4.11 What is the difference between discrete and continuous probability distributions?

Q.4.12 What type of probability distribution best models: (a) coin flip, (b) the maximum speed of a car, (c) the number of stars in the sky, (d) the volume of clouds.

Q.4.13 Assuming a certain type of cellular membrane channel has a 30% chance of being open, that there are ten channels in a given area, and that the channel's behavior follows a binomial distribution, what is the probability that:

(a) All ten will be open?
(b) All ten will be closed?
(c) No more than five will be open?

Q.4.14 How would a mutation that increased the chances of the channel being open to 40% alter the probabilities for the channel described in problem #3?

Q.4.15 Assuming a neuron has a basal rate of 3 excitations per second and the excitation rate follows a Poisson distribution; what is the probability that:

(a) No excitations will occur in one second?
(b) Less than five excitations will occur in one second?
(c) At least five excitations will occur in one second?

Q.4.16 How would a disease that reduces the basal rate of the nerve in Problem 4.15 to 2 excitations per second impact the probabilities?

Q.4.17 Protein A has a lifetime of 4 h while protein B has a lifetime of 2 h. If you begin with twice as much B as A and their decay follows an exponential distribution, which protein will almost completely disappear first?

Q.4.18 Hand draw a standard normal distribution. Assuming the mean remains 0, redraw the curve with a standard deviation of 0.5 and a second curve with a standard deviation of 2. What do you expect if the standard deviation becomes very small or very large?

4.2 Answers

A.4.1 A base 4 or 5 system could be of value in working in the context of DNA and RNA because of the fact that there are five nucleotides that are involved in the structure and of these biomolecules. This might argue for a base 5 system overall. DNA and RNA each use only 4 of the five purine or pyrimidine bases so a base 4 system might be considered.

A.4.2 This is a case of sampling with replacement. There are 150 successive boxes or slots, each of which can take one of four possible values or states, thus: $= 4^{150} = 2.037035976334487 \times 10^{90}$.

A.4.3 To show that $\binom{n}{r} = \binom{n}{n-r}$, substitute into $\binom{n}{r} = \dfrac{n!}{r!\,(n-r)!}$

$(n-r)$ *for* (r) which allows the following to be written:

$$\binom{n}{(r)} = \frac{n!}{r!\,(n-r)!} = \frac{n!}{(n-r)!\,(n-(n-r))!} = \frac{n!}{(n-r)!\,(r)!}.$$

This is the result that is desired.

A.4.4 To show that $\binom{n}{1} = n$ we start with $\binom{n}{r} = \frac{n!}{r!(n-r)!}$ and substitute with the result that $\frac{n!}{1!(n-1)!}$ which is clearly equal to n, the result sought.

A.4.5 To show that $\binom{n}{n} = 1$. Substitute into $\binom{n}{r} = \frac{n!}{r!\,(n-r)!}$ which yields

$$\binom{n}{n} = \frac{n!}{n!\,(n-n)!}$$
$$= \frac{n!}{n!\,(0)!}$$
$$= \frac{n!}{n!} = 1$$

A.4.6

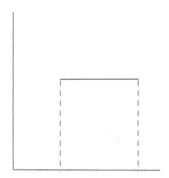

A.4.7 This problem is a case of sampling without replacement and is written and calculated as follows: $\frac{20!}{15!} = 1,860,480$

A.4.8 Total outcomes $= \binom{120}{16}$

$$P(5T, 5A, 5C, 1G) = \frac{\binom{5}{20}\binom{5}{20}\binom{5}{20}\binom{1}{15}}{\binom{120}{16}}$$

A.4.9 Use the Stirling approximation $lnN! = N\,lnN - N$ and write

$$\ln\left(1.8 \times 10^{18}\right)! = \left(1.8 \times 10^{18}\right)\ln\left(1.8 \times 10^{18}\right) - \left(1.8 \times 10^{18}\right)$$
$$= 7.386 \times 10^{19}$$

A.4.10 Prove that the binomial distribution and the Poisson distribution are nearly equivalent.

$$\binom{n}{k} p^k (1-p)^{n-k} = \frac{n(n-1)\cdots(n-k+1)}{k!} p^k (1-p)^{n-k}$$

The approach to this proof is as follows:

Use the following assumptions:

(1) n is large i.e. (there are a large number of trials)
(2) p is small i.e. (the number of successes are rare)
(3) the product λ is moderate
(4) $k \ll n$

Now rewrite the equation to be proved as follows:

$$\text{P}(k \text{successes in } n \text{ trials}) = p^k (1-p)^{n-k} \approx \frac{e^{-\lambda} \lambda^k}{k!} \qquad \text{(A.4.10.1)}$$

In this equation the term on the left of the approximation sign is the binomial and the term on the right is the Poisson. The solution requires transforming the binomial expression into the Poisson.

1. Start by expanding the binomial expression of Equation (A.4.10.1) as follows:

$$\binom{n}{k} p^k (1-p)^{n-k} = \frac{n(n-1)\cdots(n-k+1)}{k!} p^k (1-p)^{n-k} \qquad \text{(A.4.10.2)}$$

Now substitute $p = \frac{\lambda}{n}$ into (A.4.10.2) and rearrange:

$$\binom{n}{k} p^k (1-p)^{n-k} = \frac{n(n-1)\cdots(n-k+1)}{n^k} \frac{\lambda^k}{k!} \frac{(1-\lambda/n)^n}{(1-\lambda/n)^k} \qquad \text{(A.4.10.3)}$$

This expression can be simplified if the conditions described above are applied:
When n is large,

$$\left(1 - \frac{\lambda}{n}\right)^n \approx e^{-\lambda}, \quad \text{because } e^a = \lim_{x \to \infty} \left(1 + \frac{a}{x}\right)^x \qquad \text{(A.4.10.4)}$$

p is small,

$$\left(1 - \frac{\lambda}{n}\right)^n \approx 1 \qquad \text{(A.4.10.5)}$$

and $k << n$:, then the lower degree terms can be ignored,

$$\frac{n(n-1)(n-1)...(n-k+1)}{n^k} = \frac{n^k + \text{lower degree terms}}{n^k} \approx 1.$$
(A.4.10.6)

Thus A.4.10.3 can be written:

$$\binom{n}{k} p^k (1-p)^{n-k} \approx (1) \quad \frac{\lambda^k}{k!} \frac{e^{-\lambda}}{1} \approx \frac{e^{-\lambda}\lambda^k}{k!}.$$
(A.4.10.7)

This is the result that proves the case.

A.4.11 Discrete probability distributions model systems with finite, or countably infinite, values, while a continuous probability distribution model systems with infinite possible values within a range.

A.4.12 (a) discrete (Head or tails.), (b) continuous (The car's maximum speed can take on any final value within the range of possibilities.), (c) discrete (There aren't any half stars, so the number of stars must be an integer, making them "countable" in the mathematical sense."), (d) continuous (Any volume is possible.)

A.4.13 Using $\dfrac{n!}{k!(n-k)!} p^k q^{n-k}$.

$$p = 0.3, q = 0.7, n = 10. \text{ (a) } k = 10; \ 5.9 \times 10^{-6}, \text{ (b) } k = 0; \ 0.28,$$
$$\text{(c) } P(k=0) + P(k=1) + P(k=2) + P(k=3)$$
$$+P(k=4) + P(k=5) = 0.95$$

A.4.14 Using $\dfrac{n!}{k!(n-k)!} p^k q^n - k$.

$$p = 0.4, q = 0.6, n = 10. \text{ (a) } k = 10; \ 1.1 \times 10^{-4},$$
$$\text{(b) } k = 0; \ 6.0 \times 10^{-3}, \text{ (c) } P(k=0) + P(k=1)$$
$$+P(k=2) + P(k=3) + P(k=4) + P(k=5) = 0.83$$

A.4.15 Using P(outcome is k) $= \dfrac{e^{-\lambda}\lambda^k}{k!}$.

$$\lambda = 3. \text{ (a) } k = 0; 0.050, \text{ (b) } P(k=0) + P(k=1) + P(k=2)$$
$$+P(k=3) + P(k=4) = 0.82, \text{ (c) } 1 - (P(k=0) + P(k=1)$$
$$+P(k=2) + P(k=3) + P(k=4)) = 0.18$$

A.4.16   Using P(outcome is k) $= \dfrac{e^{-\lambda}\lambda^k}{k!}$.

$\lambda = 2$. (a) $k = 0$; 0.14, (b) $P(k = 0) + P(k = 1) + P(k = 2)$

$+ P(k = 3) + P(k = 4) = 0.98$, (c) $1 - (P(k = 0) + P(k = 1)$

$+ P(k = 2) + P(k = 3) + P(k = 4)) = 0.02$

A.4.17   To identify which one will disappear first, solve for the time required to reach 0.01 of one's original concentration. $\lambda_A = 4$ h; $\lambda_B = 2$ h; A_0 is the initial concentration of Protein A. Protein A: $0.01\ A_0 = A_0 \times e^{-t/4}$; $t = 18.4$ h. Protein B: $B_0 = 2A_0$; $0.01 \times 2A_0 = 2A_0 \times e^{-t/2}$; $t = 7.8\,h$. Protein B has a much shorter lifetime so it will almost completely disappear (0.01 < original amount) first.

A.4.18   See the following graphs for the appearance of different normal distribution graphs. Very small standard deviations will produce very narrow curves, while very large standard deviations will produce almost flat curves.

$\sigma = 1$

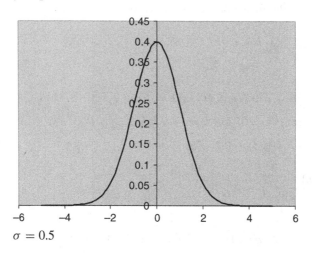

$\sigma = 0.5$

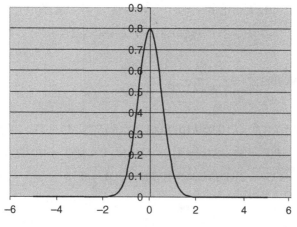

$\sigma = 2$:

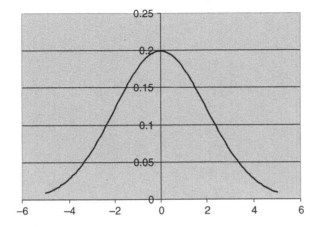

Part II
Foundations

Chapter 5
Physical Principles:
Energy – The Prime Observable

5.1 Questions

Q.5.1 Define a conservative system and give one example.

Q.5.2 (a) State the law of conservation of energy (b) If a frictionless pendulum has 4 J of energy at the top of its swing, how many Joules of energy will it have at the bottom of its swing? (Assume the pendulum is an isolated system.)

Q.5.3 Our sun's lifespan is estimated to be 10,000,000,000 years. Express the minimum lifetime of an electron in sun lifespans

Q.5.4 Which three laws of conservation mean that experimental results are independent of experiment's location in space-time? Why is this important?

Q.5.5 A 1000-kg car accelerates at a rate of 5 m/s^2 for 5 S. (a) What is the force acting on it during the first second? (b) What is its final velocity? (c) What is its final momentum?

Q.5.6 What if the same force that was applied to the car in #5 is applied to a 100 kg human for 1 S? What will be the human's (a) acceleration, (b) final velocity, and c) final momentum?

Q.5.7 If somebody uses 4 J to lift a 0.1 kg mass on Earth ($g = 9.8$ m/s^2), how much higher is the mass? How much energy would it require to lift the same mass the same distance on the moon ($g = 1.6$ m/s^2).

Q.5.8 A certain person's lungs change from 5 L to 4.5 L when they exhale. Assuming breathing can be modeled using ideal Pressure-Volume work, approximately how much work does the person do with each exhalation? (Atmospheric pressure is 101.325 kPa.)

Q.5.9 Calculate the work performed by a battery delivering 100 milliamps at 9 volts for 2 h. How much energy is used? Express the answer in (a) joules and joules-sec^{-1}, (b) calories-calories and (c) watt-hours and watts. Label which units are work and which are energy.

This chapter from *The Physical Basis of Biochemistry: Solutions Manual to the Second Edition* corresponds to Chapter 6 from *The Physical Basis of Biochemistry, Second Edition*

P.R. Bergethon, K. Hallock, *The Physical Basis of Biochemistry*,
DOI 10.1007/978-1-4419-7364-1_5, © Springer Science+Business Media, LLC 2011

Q.5.10 What is the

(a) work performed by a 50 kg gymnast who performs a lift on the rings and is elevated 1.2 m in the air?

(b) How many joules and how many calories must be supplied to the muscles?

(c) Assuming 100% efficiency in energy extraction from sugar and given that there are 4 kcal/gram of sugar and 5 g in a teaspoon, how many teaspoons of sugar should be consumed at lunch to make the lift?

5.2 Answers

A.5.1 A conservative system is one in which the energy of a point in state space is related to its position. Gravity and electrical potential are both examples of conservative forces.

A.5.2 (a) The total energy of a system is fixed and equal to the sum of the kinetic and potential energy in the system. (b) Because energy is conserved, the frictionless pendulum will have 4 J of energy at the bottom of its swing, where all of the energy will be kinetic.

A.5.3 $10^{21}/10^{10} = 100,000,000,000$ lifespans. N.B., the stability of the electron can be approximated as being stable for the period of time of interest to human survival in our solar system.

A.5.4 Mass and energy, linear momentum, and angular momentum. If experimental results depended on space-time location, reproducing experiments would be uncertain.

A.5.5 As shown in Equation 6.3 and its associated text, the force acting on an object is proportional to its mass and acceleration. Its momentum is proportional to its mass and velocity. a) $F = ma$; $F = 1000 \text{ kg} \times 5 \text{ m/s}^2 = 5000 \text{ N}$, (b) $5 \text{ s} \times 5 \text{ m/s}^2 = 25 \text{ m/s}$, (c) Momentum $= mv = 1000 \text{ kg} \times 25 \text{ m/s} = 25,000 \text{ kg} \cdot \text{m/s}$

A.5.6 As shown in Equation 6.3 and its associated text, the force acting on an object is proportional to mass and acceleration, and its momentum is proportional to its mass and velocity. (a) $F = ma$; $F/m = a$; $5000\text{N}/100\text{kg} = 50 \text{ m/s}^2$, (b) $1 \text{ s} \times 50 \text{ m/s}^2 = 50 \text{ m/s}$, (c) Momentum $= mv = 50 \text{ m/s} \times 100 \text{ kg} = 5000 \text{ kg} \cdot \text{m/s}$

A.5.7 As shown in Equation 6.14 and its associated text, work done on an object in a gravitational field is proportional to its mass, the acceleration due to gravity, and the change in its height. Personal experience tells you that it requires work to raise an object, while falling objects can be used to do work. This is why when the change in height is positive, the work done on the object is also positive. (a) $w = mg\Delta h$ can be rearranged to $w/mg = \Delta h$. $4 \text{ J}/(9.8 \text{ m/s}^2 \times 0.1 \text{ kg}) = 4.08 \text{ m}$; (b) $w = mg\Delta h = 1.6 \text{ m/s}^2 \times 0.1 \text{ kg} \times 4.08 \text{ m} = 0.65 \text{ J}$.

A.5.8 As shown in Equation 6.13 and its associated text, work on or by a change in volume under constant external pressure is proportional to the external pressure and the volume change. Because we live in a relatively constant pressure environment, this type of work is very important to biology and technology. $w = -P_{ext}\Delta V = -1.01325 \times 10^5$ Pa $\times$ (4.5 $-$ 5) liters $\times$ 0.001m^3/L $= 50.66$ J. (Note: 1 J $= 1$ Pa $\cdot$ m^3.)

A.5.9 Calculate work the using the formula, $w = -EIt$. The power delivered by the battery is $w = -EI$. The power delivered by the battery is 0.9 W or 0.9 J/S and 1.8 W h is the work performed. To perform 1.8 W h of work 1.8 W h of energy is required. Making the proper conversions gives the following:

	Energy	Power	Work
(a)	6,480 J	0.9 J S^{-1}	6,480 J
(b)	1,548 Cal	12.9 Cal S^{-1}	1,548 Cal
(c)	1.8 W h	0.9 W	1.8 w h

A.5.10 Calculate using the formula, $w = mg\Delta h$: (a) -588.4 J; (b) 588.4 J and 140 calories must be supplied to the muscles; (c) There are 16,756 J in a gram of sugar or 83.8 kJ per teaspoon of sugar. Only 35 mg of sugar or a little less than a pinch of table sugar is needed for this lift.

Chapter 6
Biophysical Forces in Molecular Systems

6.1 Questions

Q.6.1 Scuba divers will often inflate an inflexible canvas buoy underwater with gas released from their low pressure regulator mouthpiece and allow it to float to the surface in order to mark their region of ascent. A canvas bag of 3 l volume is inflated with gas bubbled into it at 60 ft below the surface.

   (a) What is the air pressure in the inflated bag after it is inflated at 60 ft depth? The water temperature at 60 ft is 20°C.
   (b) What is the pressure in the bag when it reaches the surface? The water temperature at sea level is 30°C.

Q.6.2 A new company sells a scuba marker buoy that is much smaller to carry because it has a highly flexible and elastic wall. A dive master on a similar dive to the one in problem Q.6.1 inflates the flexible buoy to 2 l using the high pressure hose from her gas cylinder. She attaches the buoy to a line and lets it go to the surface. The temperature at 60 ft is 15°C

   (a) What is the volume of the buoy at the surface if the surface water is cold and the temperature is 15°C?
   (b) What is the volume of the buoy at the surface if the surface water is warmer and the temperature is 30°C?

Q.6.3 The instructions for the new flexible buoy are emphatic that the volume at maximum capacity is 10 l. Beyond this volume the buoy will burst. A diver at 60 ft., where it is 15°C, fills the buoy with 4.2 l air and releases it. A short time later, the line that had been rising with the buoy goes limp and falls back to the diver with the broken buoy.

   (a) What height did the buoy reach before it exploded if the surface temperature is 15°C?

This chapter from *The Physical Basis of Biochemistry: Solutions Manual to the Second Edition* corresponds to Chapter 7 from *The Physical Basis of Biochemistry, Second Edition*

P.R. Bergethon, K. Hallock, *The Physical Basis of Biochemistry*, DOI 10.1007/978-1-4419-7364-1_6, © Springer Science+Business Media, LLC 2011

(b) What height did the buoy reach before it exploded if the surface temperature is 30°C?

Q.6.4 (a) Calculate the proportionality coefficient of the ideal gas law, R, the gas constant from the ideal gas law using STP. (b) Express R in units of J K^{-1} mol^{-1}.

Q.6.5 The following equation is the fundamental biochemical energy reaction in heterotrophic life:

$$C_6H_{12}O_6 \text{ (glucose)} + 6O_2 \rightarrow 6CO_2 + 6H_2O$$

one liter of 1 M glucose (MM = 180.16 g/mole, density = 1.54 g/mL) is added to a sealed 6 l reaction chamber that contains 6 moles of oxygen. A nominal amount of enzyme to catalyze the above reaction is in the glucose solution. The initial temperature of the chamber is 300 K. The complete reaction of the glucose and oxygen proceeds according to the reaction, the final temperature of the chamber is 380 K.

(a) What is the initial pressure in the chamber?
(b) What is the final pressure in the chamber?

Q.6.6 Examine Table 6.Q.6.6 for boiling points and the coefficient a of the van der Waals gas equation for various substances. Comment on the relationship and what it implies about the coefficient a.

Table 6.Q.6.6   van der Waals coefficients for various substances

Substance	a(atm L^{2}mol^{-2})	b(L mol^{-1})	boiling point	mol radius
He	0.0341	0.0238	4.22	140 pm
H$_2$	0.242	0.02651	20.2	120 pm
Ar	1.335	0.03201	87.3	188 pm
N$_2$	1.350	0.0387	77.36	155 pm
O$_2$	1.362	0.03186	90.2	152 pm
Kr	2.292	0.0396	119.93	202 pm
Hg	5.120	0.01057	629.88	155 pm
H$_2$O	5.459	0.03049	373.15	275 pm

Q.6.7 Use the van der Waals equation to determine the molecular size of a gas atom of helium, neon and water vapor.

Q.6.8 For an electron with mass of 9.11×10^{-31} kg placed in a field generated by a pair of two oppositely charged parallel plates with $\sigma = 1.0 \times 10^{-6}$ C/m^2 and separated by 1 cm, (a) what is the acceleration caused by the electric field? (b) what is its velocity when it reaches the opposite plate? (c) how long will it take to reach the plate?

6.2 Answers

A.6.1 For simplicity, we will assume the gas behaves as an ideal gas.

  (a) The scuba diver takes the water-filled container and fills it with air, driving out the water. While filling the bag, the air pressure in the bag will equal the water pressure outside the bag. A good approximation of the additional pressure from water is that 30 ft of water increases the pressure by 1 atm, so at 60 ft, the pressure in the bag will be 3 atm. (Two atmospheres of pressure from water and one atmosphere of pressure from that air above the water.)

  (b) We begin with the ideal gas law. $PV = nRT$. Because the container is inflexible and seals, V and n will be the same regardless of depth, so we rearrange the above equation into: $P_1/T_1 = P_2/T_2$. $P_1 = 3$ atm, $T_1 = 293$ K, $T_2 = 303$ K. Rearranging and solving for $P_2 = T_2 \times P_1/T_1 = 3$ atm $(303 \text{ K}/293 \text{ K}) = 3.1$ atm.

A.6.2 (a) We begin with the ideal gas law. $PV = nRT$. Because the container is flexible and the temperature on the surface is the same as underwater, T and n will be the same regardless of depth, so we rearrange ideal gas law into: $P_1V_1 = P_2V_2$. $P_1 = 3$ atm, $V_1 = 2$ L, $P_2 = 1$ atm. Rearranging and solving for $V_2 = P_1V_1/P_2 = 21 \times (3 \text{ atm}/1 \text{ atm}) = 6$ l.

  (b) Again we begin with the ideal gas law. $PV = nRT$. This time, the only possible variable that remains constant is n because the container is sealed. We rearrange ideal gas law into: $P_1V_1/T_1 = P_2V_2/T_2$, $P_1 = 3$ atm, $V_1 = 2$ L, $T_1 = 288$ K, $P_2 = 1$ atm, $T_2 = 303$ K,. Rearranging and solving for $V_2 = P_1V_1T_2/T_1P_2 = (21 \times 3 \text{ atm} \times 303\text{K})/(1 \text{ atm} \times 288\text{K}) = 6.3$ l.

A.6.3 (a) We begin with the ideal gas law. $PV = nRT$. Because the container is flexible and the temperature on the surface is the same as underwater, T and n will be the same regardless of depth, so we rearrange ideal gas law into: $P_1V_1 = P_2V_2$. We want to know at what depth the buoy pops, which will be when its volume exceeds 10 l, so the first step is to find the pressure at which the volume of the buoy is 10 l. $P_1 = 3$ atm, $V_1 = 4.2$l, $V_2 = 10$l. Rearranging the ideal gas law, we have $P_2 = P_1V_1/V_2 = 3$ atm $\times (4.2 \text{ l}/10 \text{ l}) = 1.26$ atm. Now we have to find the depth, which can be found by using the relation described in problem 6.1 30 ft water $= 1$ atm. The pressure due to water is 1.26 atm $- 1$ atm $= 0.26$ atm. 0.26 atm $\times 30$ ft water/1 atm $= 7.8$ ft. The buoy will burst at a depth of 7.8 ft.

  (b) To determine this, we'll need to know the change in volume due to both pressure and temperature changes as the buoy ascends. Volume changes due to pressure are almost instantaneous, but as everyone's personal experience will attest, changes in temperature take considerably more time. It's reasonable to approximate that the air inside the buoy does not have time to equilibrate with the surrounding water's temperature so that its temperature remains constant until it bursts at a depth of 7.8 ft. The ideal gas law assume everything is at equilibrium, but as we will discuss later, sometimes reactions move too fast for this to be a valid approximation.

A.6.4 (a) $R = \dfrac{(1\,\text{atm})\,(22.414\,\text{L})}{(1\,\text{mol})\,(273.15\,\text{K})} = 0.08206\,\text{L atm K}^{-1}\,\text{mol}^{-1}$

(b) $R = \dfrac{(1.01325\times10^{5}\text{N m}^{-2})\,(22.414\times10^{-3}\text{m}^{3})}{(1\,\text{mol})\,(273.15\,\text{K})} = 8.314\,\text{N m K}^{-1}\,\text{mol}^{-1}$

therefore $0.08206\,\text{L atm K}^{-1}\,\text{mol}^{-1} = 8.314\,\text{JK}^{-1}\,\text{mol}^{-1}$

A.6.5 (a) To calculate the initial pressure, we'll assume oxygen behaves as an ideal gas, it is insoluble in water, and water is an incompressible liquid. With these assumptions, we have 6 moles of oxygen in 5 l of space at 300 K. $PV = nRT$ can be rearranged into $P = nRT/V = (6\ \text{moles} \times 0.0821\ \text{atm l mole}^{-1}\,\text{K}^{-1} \times 300\ \text{K})/5\ \text{l} = 29.56$ atm.

(b) If we assume that CO_2 behaves as an ideal gas, the 6 moles of oxygen that is consumed in the reaction is replaced by 6 mole of CO_2, so there isn't any pressure change due to a change in gas volume. The glucose is consumed in the reaction, which assuming ideal mixing, the decrease in volume due to the loss of glucose is:

1 mole/l × 1 l × 180.16 g/mole × 1 ml/1.54 g = 117 ml

The volume of solution decreases 117 ml due to the loss glucose, but it increases when 6 moles of water is created.

6 moles water × 18 g/mole × 1 g/ml = 108 ml.

So the final solution volume is 1000 ml − 117 ml + 108 ml = 991 ml = 0.991 l

The final volume the gas occupies is 6 l − 0.991 l = 5.009 l and its final temperature is 380 K.

$P = nRT/V = (6\ \text{moles} \times 0.0821\ \text{atm l mole}^{-1}\,\text{K}^{-1}$
$\times 380\ \text{K})/5.009\ \text{l} = 37.37\ \text{atm}.$

A.6.6 There is a clear trend that the coefficient a and the boiling point are positively correlated. The internal energy required before a substance will change state from the more highly associated liquid state into the vapor phase is determined by the force of attraction between the molecules. Thus the observed relationship is consistent with fact that a corrects the ideal gas equation for the force of attraction between gas particles, this force is ignored in the ideal gas law but accounted for in the van der Waals treatment. Consistent with this interpretation, gases with very small values of a, such as H_2 and He, are cooled to almost absolute zero before they condense to form a liquid.

A.6.7 The constant b is an indication of molecular volume, thus it could be used to estimate the radius of an atom or molecule, modeled as a sphere. The value of b in Table 6.Q.6.6 indicates that a mole of krypton gas occupies 0.0396 l. This can be used to estimate the radius: 6.02×10^{23} krypton spheres will occupy the volume given by b. A single argon particle will therefore occupy a volume:

$$V_{Ar} = \frac{39.6 \times 10^{-6} m^3 \times mol^{-1}}{6.02 \times 10^{23} particle \times mol^{-1}}$$

$$= 6.579 \times 10^{-29} m^3 \times particle^{-1}$$

The radius of this argon sphere can be calculated using the formula for the volume of a sphere: $V = 4/3\pi r^3$. Substituting and solving for r gives:

$$6.579 \times 10^{-29} m^3 \times krypton^{-1} = 4/3\pi r^3$$
$$2.504 \times 10^{-10} = r_{krypton}$$

This is value of 250 pm is close but not completely in agreement with the measured crystallographic radius of krypton.

A.6.8 The acceleration of the electron is calculated first:

$$\mathbf{a} = \frac{q_{ext}}{m} = \frac{q_{ext}\mathbf{E}}{m} = \frac{q_{ext}\sigma}{m\varepsilon_o}$$

$$= \frac{(1.6 \times 10^{-19} \text{ C})(1.0 \times 10^{-6} \text{ C/m}^2)}{(9.11 \times 10^{-31} \text{ kg})(7.85 \times 10^{-12} \text{ C}^2/\text{N} - \text{m}^2)}$$

$$= 4.9845 \times 10^{16} \text{ m/s}^2$$

Since the electron can travel 1 cm from one plate before striking the other plate, it will attain a velocity of

$$v^2 = 2ax$$
$$v = \left[2(1.9845 \times 10^{16} m/s^2)(1 \times 10^{-2} \text{ m})\right]^{1/2}$$
$$= 1.992 \times 10^7 m/s$$

The time to make the crossing between the plates is:

$$t = \frac{v}{a} = \frac{1.992 \times 10^7 \ m/s}{1.9845 \times 10^{16} \ m/s^2}$$
$$= 1.003 \times 10^{-9} s$$

Chapter 7
An Introduction to Quantum Mechanics

7.1 Questions

Q.7.1 The Earth's sun has a surface temperature of 5800 K. What is its color?

Q.7.2 Determine the effect of altering the temperature of a star (or going to another star system) on the distribution of light and its effect on photosynthesis given the absorption spectrum of chlorophyll vs. rhodopsin vs. retinol.

Q.7.3 (a) Show that $e^{hv/kT} ---> 1+hv/kT$ for the condition $hv/kT ---> 0$.
        (b) Show that $e^{hv/kT} ---> \infty$ for the condition $hv/kT ---> \infty$.

Q.7.4 Show that (a) the Rayleigh-Jeans law is a special case of Planck distribution law for the blackbody spectrum. Show also that (b) the Wein displacement law can be derived from Planck's distribution law.

Q.7.5 At what mass and length would a pendulum have to be constructed to reach the level of demonstrating a discontinuous energy loss?

Q.7.6 Which biologically important entities should be treated as quantized (Hint consider the size and speed of the entity and apply the deBroglie relationship and then calculate $\Delta E/E$).

Q.7.7 There is a very striking and significant conclusion that must be drawn about the wave theory of light from the behavior of the photoelectric effect. What is it?

Q.7.8 Explain the cause of Fraunhöfer lines.

Q.7.9 What two words succinctly describe classical mechanics' fundamental view of state space?

Q.7.10 What was the ultraviolet catastrophe and how did Planck's proposal resolve it?

Q.7.11 How did Einstein explain Lenard's observations that the maximum kinetic energy of photoelectrons depended on the frequency of light and not the intensity?

Q.7.12 What is DeBroglie's postulate?

This chapter from *The Physical Basis of Biochemistry: Solutions Manual to the Second Edition* corresponds to Chapter 8 from *The Physical Basis of Biochemistry, Second Edition*

P.R. Bergethon, K. Hallock, *The Physical Basis of Biochemistry*, DOI 10.1007/978-1-4419-7364-1_7, © Springer Science+Business Media, LLC 2011

Q.7.13 What are the complementary observables for the Heisenberg indeterminacy or uncertainty principle? How are the complementary observables related to each other?

Q.7.14 What is tunneling? If you walk into a door, is there any chance you will tunnel through it? What is the probability?

Q.7.15 What wavelength does the Balmer series predict for $n = 3$ and $n = 4$ hydrogen lines?

Q.7.16 Calculate the deBroglie wavelength of both a 100 kg person and 14 kg dog traveling at (1) 5 m/s, (2) 50 m/s, (3) 500 m/s, (4) 5000 m/s, (5) 50,000 m/s, (6) 300,000 m/s.

Q.7.17 An electron microscope is designed with an accelerating potential of 200 kV. What is the wavelength of the electron beam in this device? What is the theoretical resolution of this microscope? Use an angle of illumination of 4×10^{-3} radians.

7.2 Answers

A.7.1 To determine the color, we have to determine the wavelengths the surface emits, which can be done by calculate its maximum emission wavelength (λ_{max}). $\lambda_{max} \times T = 2.898 \times 10^{-3}$ m K. Since the sun's surface temperature is 5800 K, we rearrange the equation and get $\lambda_{max} = 2.898 \times 10^{-3}$ m K/5800 K $= 500$ nm. 500 nm is in the green-blue portion of the spectrum, so it would be the dominant color.

A.7.2 If the temperature of the star increases or decreases, the peak wavelength will increase or decrease in wavelength, respectively. Plants would have to shift their frequencies of activity as well to make use of both edges of the spectrum. However, there are energetic constraints on the wavelengths that can be efficiently used for electron transport so if the temperature of the star changes too much, chlorophylls may have to use wavelengths that are closer together. Lifeforms that use rhodopsin and retinol for light detection might need to have those modified so they can still see, depending on how much the spectrum shifts.

A.7.3 (a) Write $h\nu/kT = z$ and let $z = 1, 0.75.0.5, 0.25, 0.15, 0.1,$ and 0. Calculating e^z and comparing the results with the sums of $1 + z$ (in parentheses) gives the series of numbers 2.718 (2), 2.117 (1.75), 1.648 (1.5), 1.28 (1.25), 1.16 (1.15), 1.105 (1.1), and 1 (1).

(b) Write $h\nu/kT = z$ and let $z = 1, 10. 100,$. Calculating e^z and comparing the results with the value of z (in parentheses) gives the series of numbers 2.718 (1), $2.20 \times 10^4 (10)$, and $2.69 \times 10^{43} (100)$. Even with just three calculations, the trend is clear. Small increases in z substantially increase e^z.

A.7.4 Show that (a) the Rayleigh-Jeans law is a special case of Planck distribution law for the blackbody spectrum. Show also that (b) Stefan's law and (c) the Wein displacement law can be derived from Planck's distribution law.

Planck's distribution law is $\rho_T (v)\, dv = \dfrac{8\pi v^2}{c^3} \dfrac{hv}{e^{hv/kT} - 1} dv$

(a) As shown in question 7.3, $e^{hv/kT} - - - > 1 + hv/kT$ when $hv/kT - - - > 0$, so at low frequencies, the second term becomes $hv/(1 + hv/kT - 1) = kT$. Replacing the second term with kT gives the Rayleigh-Jeans law. $\rho_{(v.T)}dv = \dfrac{8\pi v^2 kT}{c^3} dv$

(b) To find the maximum frequency at a given temperature, we will need to take a derivative of Planck's law with respect to frequency and set it equal to 0; $d\rho/dv = 0$. At x values, $(e^x - 1) \approx e^x$, so at high frequencies, $1/(e^{hv/kT} - 1) \approx e^{-hv/kT}$. Rearranging Planck's distribution law, collecting all of the constants into one value, C, and using the high frequency approximation, we get: $\rho_{(v,T)} = Cv^3 e^{-hv/kT}$.

$$d\rho/dv = 0 = d(Cv^3 e^{-hv/kT})/dv = 3 \times Cv^2 e^{-hv/kT}$$

$$-(h/kT)Cv^3 e^{-hv/kT}, \text{rearranging,}$$

$3{*}Cv^2 e^{-hv/kT} = (h/kT)Cv^3 e^{-hv/kT}$, divide the common terms from both sides

$3 = (h/kT)v$, rearranging,

$v = 3kT/h$, which is consistent with Wein's displacement law.

A.7.5 The pendulum would need to be on the order of a molecule, so using water as a model, its mass would need to near 10^{-26} kg and its length in the nanometer range.

A.7.6 Electrons and protons are the most important biological entities that require quantum treatment on a routine basis.

A.7.7 The photoelectric effect implies that the entire quantum of energy is imparted to a single electron. This means that the *size* of the light quantum is quite small and must be confined to a very small region of space. This tightly packed bundle of energy can not be a wave because a wave would have the energy evenly spread across its entire front and the energy necessary to eject an electron would accumulate slowly. The observation that emission is an instantaneous process supports the description of the light quantum as a packet of energy or photon and not as a wave. A light beam is thus a stream of corpuscles as Newton had first suggested.

A.7.8 The dark lines in the solar spectrum are caused by elements in the sun's atmosphere that absorb at those specific wavelengths. The existence of helium was demonstrated in the Sun's atmosphere before it was found on the Earth. The dark lines of helium in the solar spectrum had the same pattern as the emission spectrum of elemental helium discovered on the Earth.

A.7.9 Determinism and continuum.

A.7.10 The Rayleigh-Jeans formulation worked at low frequencies of radiation, but at higher frequencies it predicted infinite energy densities. Since infinite energy densities are impossible and classical treatments could not resolve this problem, it was dubbed the ultraviolet catastrophe. Planck rejected the assignment of the mean energy for all standing waves as given by equipartition theory, replacing it with a postulate that every energy state was distinct and each harmonic oscillation was an energy step that's proportional to the frequency. Planck's theory modeled the low and high frequency black-body radiators better than Rayleigh-Jeans.

A.7.11 Einstein proposed that light energy is carried in discreet packets and that only those packets with enough individual energy can remove an electron. Any additional energy above the needed energy is converted into the photoelectron's kinetic energy.

A.7.12 There is a relationship between the momentum of a particle and its wavelength. $p = h/\lambda$.

A.7.13 Momentum and position; time and energy. The complementary observables cannot be both known with an arbitrarily high degree of accuracy. For example, one cannot know exactly the position and momentum of a particle.

A.7.14 Tunneling is the finite probability that a particle will be transmitted through a barrier into a forbidden zone, e.g. a low-energy electron might pass through a high energy barrier even though it does not have enough energy to go over the barrier. Technically, there's a finite probability of a person tunneling through a door, but we do not recommend walking into doors hoping to tunnel through it because the probability is vanishingly small.

A.7.15 Using Equation (8.22): $\lambda = 3646 \dfrac{n^2}{n^2 - 4}$ for $n = 3, 4, 5 \ldots$

(a) 6562.8 Å, (b) 4861. 3 Å

A.7.16 Looking at Table 8.2 will provide a good estimate before any calculation. A running person probably falls within the baseball, speeding bullet, and elephant range, so one should expect a value in the 10^{-34}–10^{-36} range. The calculation requires the de Broglie relation, which is stated in the text between Equations (8.28) and (8.29). $p = h/\lambda$; rearranged to be $\lambda = h/p$. h $= 6.626 \times 10^{-34}$ J $\cdot$ s. p $= m^*v = 100$ kg*5 m/s $= 500$ kg $\cdot$ m/s. $\lambda = h/p = 6.626 \times 10^{-34}$ J $\cdot$ s/500 kg $\cdot$ m/s $= 1.33 \times 10^{-36}$ m.

100 kg human deBroglie wavelength (m) (a) 1.33×10^{-36}, (b) 1.33×10^{-37}, (c) 1.33×10^{-38}, (d) 1.33×10^{-39}, (e) 1.33×10^{-40}, (f) 1.33×10^{-41}. 14 kg dog deBroglie wavelength (m) (a) 9.47×10^{-36}, (b) 9.47×10^{-37}, (c) 9.47×10^{-38}, (d) 9.47×10^{-39}, (e) 9.47×10^{-40}, (f) 1.58×10^{-40}.

A.7.17 First calculate the deBroglie wavelength of the electron at the accelerating voltage: $\lambda = \dfrac{1.23\,nm}{V^{1/2}} = \dfrac{1.23 \times 10^{-9}}{2 \times 10^5} = 2.75 \times 10^{-12} m$. Then use this wavelength in the equation from Chapter 8: $d = \dfrac{0.61\lambda}{\alpha}$. The result is a theoretical resolution of 4.194×10^{-10} m.

Chapter 8
Chemical Principles

8.1 Questions

Q.8.1 Calculate the Coulomb interaction for Na^+ and Cl^- when separated by the distance of their atomic radii.

Q.8.2 Calculate the magnitude of the thermal energy kT at 4.2 K (boiling point of He); 77.36 K (boiling point of N_2); 90.2 K (boiling point of O_2); 298 K (room temperature); 310 K (body temperature); at 373 K (boiling point of water) ?

Q.8.3 Compare the magnitude of the energy of the Coulomb attraction from question Q.8.1 with the thermal energy at room temperature (298 K).

Q.8.4 What separation is needed to make the Coulomb attraction energy approach the scale of the thermal energy at room temperature?

Q.8.5 What is the energy U(r) and force F(r) predicted by the Lennard-Jones potential when two atoms are at their equilibrium (minimum) separation. $A = 10^{-77} J\,m^6 ; B = 10^{-134} J\,m^{12}$.

Q.8.6 Mammals have very tightly controlled temperature ranges compatible with life. Temperatures below 27°C and above 41°C generally result in death probably from loss of activity of enyzme activity and normal receptor-agonist binding. Assume the major contribution to these binding interactions are dispersion forces and hydrogen bonds. Further assume that the conformation of the enzymes and receptors does not change significantly at these temperatures. Is the variation in thermal energy likely to change the energy of interaction significantly?

Q.8.7 What are some of the problems in computation that would occur if the Born-Oppenheimer approximation were not used?

This chapter from *The Physical Basis of Biochemistry: Solutions Manual to the Second Edition* corresponds to Chapter 9 from *The Physical Basis of Biochemistry, Second Edition*

P.R. Bergethon, K. Hallock, *The Physical Basis of Biochemistry*,
DOI 10.1007/978-1-4419-7364-1_8, © Springer Science+Business Media, LLC 2011

8.2 Answers

A.8.1   The energy of interaction between these two atoms is found using Coulomb's law: $U_{i-i} = k\frac{q_{Cl} q_{Na}}{r}$. $k = \frac{1}{4\pi\varepsilon_o} = 8.988 \times 10^9 \text{N} \times \text{m}^2 \times \text{c}^{-2}$. The atomic radii of Na^+ is 0.095 nm and Cl^- is 0.181 nm thus the interatomic radius is 0.276 nm.
Substituting this

$$U_{i-i} = k\frac{q_{Cl} q_{Na}}{r}$$

$$= 8.988 \times 10^9 \text{N} \times \text{m}^2 \times \text{c}^{-2}\frac{(+1.602 \times 10^{-19}\text{c})(-1.602 \times 10^{-19}\text{c})}{.276 \times 10^{+9}\text{m}}$$

$$= 8.358 \times 10^{-19}\text{N} \times \text{m}$$

A.8.2   All of these are calculated by multiplying the Boltzmann constant by the temperature: $1.3806503 \times 10^{-23}\text{m}^2 \text{ kg s}^{-2} \text{ K}^{-1}$(T)

at 4.2 K (boiling point of He); 5.7988×10^{-23} J
77.36 K (boiling point of N_2); 1.0680×10^{-21} J
90.2 K (boiling point of O_2); 1.2453×10^{-21} J
298 K (room temperature); 4.1143×10^{-21} J
310 K (body temperature); 4.2800×10^{-21} J
373 K (boiling point of water) 514.9825×10^{-23} J

A.8.3   At room temperature the electrostatic interaction energy is approximately 200 times greater in magnitude than the thermal energy,

Ion-ion v.s. thermal: -4.1×10^{-21} J v.s. 8.358×10^{-19} J

A.8.4   At 56 nm separation the attractive electrostatic interaction and the randomizing thermal energy will be essentially equal.

A.8.5   U(r) is minimum at $\frac{dU}{dr} = 0$.

This occurs when $r_{equilibrium} = \left(\frac{2B}{A}\right)^{1/6}$

Thus $U_{equilibrium} = \frac{-A^2}{4B} = -\frac{A}{2r_{equilibrium}^6} = -2.5 \times 10^{-21}$ J.

$$F = -\frac{dU}{dr}; F_{max} = \frac{d^2U}{dr^2} = 0, \text{ which occurs when } r_{separation}$$

$$= \left(\frac{26B}{7A}\right)^{1/6} = 0.3935 \text{ nm}$$

$$F_{max} = \frac{6A}{r^7} + \frac{12B}{r^{13}} = \frac{126A^2}{169B}\Big/\left(\frac{26B}{7A}\right)^{1/6} = -1.89 \times 10^{-11}\text{N}$$

A.8.6 The difference in thermal energy at 300 K and 314 K is 14 K or 1.933×10^{-22} J. Compared to the 0.276 nm separation of two elementary charges this energy is about 4300 times less. The separation between charges of a Coulombic nature at an equivalent magnitude would necessarily be 1193 nm, thus such a small change in the thermal energy would be unlikely to affect most Coulombic interactions. van der Waals and hydrogen bonding interactions fall off rapidly at an r^{-6} relationship. Thus an increased distance of only 5 times or less of the interatomic distance (about 3 nm) would be equal to the energy difference represented by the 14 K differential. This is a small distance for the intermolecular interactions to have structure function tolerance so it is a reasonable argument that these forces could account for the sensitivity of enzyme and receptor specificity in this temperature range.

A.8.7 The Born-Oppenheimer approximation recognizes that the analytic computation of the many-body problem that is represented in the electronic and nuclear structure is not computationally tractable. The approximation is essential to simplifying the otherwise unsolvable mathematical problem.

Chapter 9
Measuring the Energy of a System: Energetics and the First Law of Thermodynamics

9.1 Questions

Q.9.1 Define the First Law of Thermodynamics

Q.9.2 Define a thermodynamic system, boundary, and surroundings.

Q.9.3 Provide two examples of intensive properties and two examples of extensive properties.

Q.9.4 Heat capacity measures the amount of heat required to heat a certain amount of material, and specific heat capacity is the heat capacity divided by the mass of the material. Is heat capacity an intensive or extensive property? Is specific heat capacity an intensive or extensive property? Why?

Q.9.5 The heat capacity of 150 g apple is 547 J/K; what is its specific heat capacity?

Q.9.6 The specific heat capacity of an egg is 3.18 J/g·K and for water, it's 4.186 J/g·K. How much heat is required to bring a 50 g egg to boil in a 500 mL of water. (Assume the starting temperature is 20°C.)

Q.9.7 What is the change in enthalpy for the egg?

Q.9.8 An amateur geologist digs up a chunk of unknown metal while hiking through nearby mountains. After returning with the 1.2 kg chunk of rock a 10°C change is measured after only 2.9 kJ of heat is added to the metal chunk. What is the chunk's specific heat capacity?

Q.9.9 How much heat would be required to heat the metal chunk from room temperature (20°C) to 327.5°C?

Q.9.10 After reaching 327.5°C, a large section of the metal chunk melts, demonstrating that the metal chunk is a mixture of at least two different metals. A 0.8 kg dull silver chunk has a specific heat capacity of 0.13 kJ/kg·K. Can the specific heat capacity of the remaining chunk be determined without directly measuring it? If yes, calculate it.

Q.9.11 For each of the following, use systems analysis to identify (1) the system under study [defined by the elements and rules of interaction], and,

This chapter from *The Physical Basis of Biochemistry: Solutions Manual to the Second Edition* corresponds to Chapter 10 from *The Physical Basis of Biochemistry, Second Edition*

P.R. Bergethon, K. Hallock, *The Physical Basis of Biochemistry*,
DOI 10.1007/978-1-4419-7364-1_9, © Springer Science+Business Media, LLC 2011

(2) its boundary and surroundings [this defines the context or background space].

(a) cell, (b) planet earth, (c) leaf, (d) Human brain.

Q.9.12 For the following properties of a system related to *matter* identify which are intensive and which are extensive.

(a) concentration, (b) density, (c) mass.

Q.9.13 For the following properties of a system related to *PVT* identify which are intensive and which are extensive.

(a) molar volume, (b) pressure, (c) specific volume, (d) temperature, (e) volume.

Q.9.14 For the following properties of a system related to *thermal energy* identify which are intensive and which are extensive.

(a) energy, (b) entropy, (c) enthalpy, (d) heat capacity, (e) free energy, (f) specific heat (C_p/gram), (g) molar energy, (h) molar entropy, (i) molar enthalpy.

Q.9.15 For the following properties of a *system* identify which are intensive and which are extensive.

(a) area, (b) dielectric constant, (c) force, (d) mass, (e) refractive index, (f) viscosity, (g) volume, (h) zeta potential, (i) chemical potential.

Q.9.16 The energy of a system is easily measured by thermal techniques. When the state of a system and a change in the state observables is temperature dependent, an exothermic or endothermic change can be measured. For the following list of events indicate whether a physical change is endothermic or exothermic.

Adsorption	Vaporization
Desorption	Sublimation
Freezing	Dehydration
Melting	Desolvation

Q.9.17 What amount of heat is needed to change the temperature of 0.5 l of H_2O from 275 K to 325 K? The pressure is 1 atmosphere. $C_p = 1$ calg^{-1} deg^{-1}.

Q.9.18 If the same heat were added to an equivalent mass of ethanol ($C_p = 111.4$ J K^{-1} mol^{-1}) what would the temperature change of the liquid ethanol be?

Q.9.19 Using the bond energies given below calculate the heat of formation of gaseous hexane, cyclohexane and benzene. Compare your answers with the thermodynamic heats of formation given in Table 10.1. Explain any discrepancies.

Bond type	Bond dissociation energy (kJ mol^{-1})
C – C	344
C(Graphite)	716
C – H	415
C = C	615
H2	436
O2	498

Q.9.20 a) The ideal temperature for a certain refreshment is 4°C. How much ice, cooled to −20°C should be added to 250 ml of the refreshment at 25°C in order to cool it to 4°C with no dilution of the drink (i.e. no ice melts into the drink)?

b) If the drink is served in a 300 ml cup (maximum ice = 50 cc), what is the minimal amount of ice that can be added to achieve temperature and have minimal dilution? Assume the system to be adiabatic. The molar C_p of ice is 37.8 J K^{-1} mol^{-1}.

9.2 Answers

A.9.1 In differential form this is usually written, $dU = dq + dw$. Energy is conserved in every process involving a thermodynamic system and its surroundings.

A.9.2 The system is the part of the physical universe whose properties are under investigation. The boundary separates the system from the surroundings. The surroundings are everything in the physical universe not included as part of the system.

A.9.3 Temperature, density, pressure are intensive properties. Mass, volume, and entropy are extensive properties.

A.9.4 Heat capacity is an extensive property because doubling the amount of material will double the measured heat capacity. Specific heat capacity is intensive because it is independent of the amount of material present.

A.9.5 Specific heat capacity is found by dividing the mass of an object by its heat capacity. In this case, we have 547 J/K/150 g = 3.65 J/g · K

A.9.6 In this problem, we have two different components with different specific heat capacities: an egg and water. To find out how much heat is required, we calculate how much heat is required for each individual object and then add those amounts together to get the total heat required. (50 g × 3.18 J/g · K + 500 mL × 1 g/mL × 4.186 J/g · K) × (100°C − 20°C) = 180.18 kJ.

A.9.7 The change in enthalpy is calculated by multiplying the mass by the specific heat capacity, to find the total heat capacity of the egg, and then multiplying that result by the change in temperature. (50g × 3.18 J/g · K) × (100°C − 20°C) = 12.72 kJ

A.9.8 The specific heat capacity is found by dividing the total heat added to the object by the mass of the object multiplied by the change in temperature. 2.9 kJ/(1.2 kg × 10 K) = 0.24 kJ/kg · K

A.9.9 Assuming the specific heat capacity is constant with temperature, the total heat required is found by multiplying the mass by the temperature change and then by the specific heat capacity. 1.2 kg × (327.5°C − 20°C) × 0.24 kJ/kg · K = 88.56 kJ.

A.9.10 Yes, she can calculate it using the values reported in Q.9.8 (0.24 kJ/kg·K) and the measurement from the isolated 0.8 kg chunk. The process is the reverse of the process used in Q.9.6 in this chapter. The heat absorbed by the 0.8 kg chunk is first calculated, and then subtracted from the total heat. This combined with the mass of the second chunk allows the specific heat capacity for the 0.4 kg section to be calculated. Let's use the 10°C change reported in Q.9.8 for this calculations. We know the total amount of heat absorbed must equal the sum of its parts; therefore 2.9 kJ = (0.8 kg × 10 K × 0.13 kJ/kg · K + 0.4 kg × 10K × C_{p2}) = 1.04 kJ + 4 kg · K × C_{p2}). Rearranging, 1.86 kJ = 4 kg · K × C_{p2}; dividing by 4 kg · K, C_{p2} = 0.47 kJ/kg · K.

A.9.11 In each of these cases the definition of what constitutes the system in terms of its parts and their relationship, what constitutes the surroundings and how they are bounded can be a matter of discussion and debate. Reasonable simplifications are given.

System	Boundary	Surroundings
Cell (cytoplasmic and nuclear contents)	Cell membrane (including all molecules directly attached to the limiting membrane)	Extracellular spaces
Earth (planet and atmosphere contents)	1. Karman Line @ 100 km above Earth Surface 2. NASA considers 122 km above Earth Surface as where reentry by space vehicles defines boundary	Outer space
Leaf (cells, gases, vessels and chemicals)	Cuticle at leaf surface and some defined connection of the leaf to the stem of the plant.	Stem of the plant, spaces outside of the cuticle containing air, light, etc
Brain (neurons, glial cells,)	Surface of neural elements on outside, blood-brain barrier formed by glial-vascular boundary inside. There is also a "theoretical" or anatomical boundary at the base of the medulla oblongata	Cerebrospinal fluid, blood vessels, meninges, anatomically the spinal cord. The remainder of the body.

A.9.12 Properties related to matter are:
    intensive: concentration, density.
    extensive: mass.

A.9.13 Properties related to *PVT* are:

    intensive: specific volume, molar volume, pressure, temperature.
    extensive: volume.

A.9.14 Properties related to thermal energy are:

    intensive: specific heat (*Cp*/gram), molar energy, molar entropy, molar enthalpy, chemical potential.
    extensive: energy, entropy, enthalpy, free energy, heat capacity.

A.9.15 Properties related to systems are:

    intensive: chemical potential, dielectric constant, refractive index, viscosity, zeta potential
    extensive: area, force, mass, volume

A.9.16

Event	Exothermic	Endothermic
Adsorption	X	
Vaporization		X
Desorption		X
Sublimation		X
Freezing	X	
Dehydration		X
Melting		X
Desolvation		X

A.9.17 There is a 50 K change in temperature of 500 g of water. The specific heat at constant pressure is 1 cal g^{-1} deg^{-1}. The heat added is 25,000 cal or 25 kcal or 104.6 kJ.

A.9.18 104.6 kJ of heat is added to 500 g of ethanol. Calculate the moles of ethanol: $\dfrac{500g}{46.07g\ mol^{-1}} = 10.85$ mol. The heat added per mole is $\dfrac{104.6\ kJ}{10.85\ mol} =$ 9.64 kJ mol^{-1}. The temperature change is $\dfrac{9640\ J\ mol^{-1}}{111.4\ J\ K^{-1}mol^{-1}} = 86.54$ K

A.9.19 (a) *Cyclohexane:*

$$6C + 6H_2 --> C_6H_{12}$$

$$6\ C(graphite) --> 6C;\ \Delta H = 6(716.7) = 4300\ kJ$$

$6H_2 --> 12H; \quad \Delta H = 6(436) = 2616 \text{ kJ}$

$6C + 12H --> 6\,C - C + 12\,C - H; \; \Delta H = 6(-344) + 12(-415)$

$\Delta H_f = -128 \text{ kJ vs} - 123 \text{ kJ from table.}$

Hexane: -178 kJ vs $- 167$ kJ.

Benzene: 241 kJ vs 82.93 kJ. The lower thermodynamic enthalpy represents the resonance energy gained by the formation of the aromatic ring.

A.9.20 (a) Temp drop for liquid $=$ 21.97 kJ. Temp rise for ice without phase change $= 24°$C. Will need 24.22 moles of ice $= 436$ g of ice.

Chapter 10
Entropy and the Second Law of Thermodynamics

10.1 Questions

Q.10.1 Consider a three compartment container into which the same particles described in Section 11.3.2 are placed. Calculate the probability of: (a) 1 particle being found in the middle compartment, (b) 10 particles, (c) 100 particles.

Q.10.2 These particles from Problem Q.10.1 re-sort themselves every pico-second. Assuming that you have a machine to look at the compartments every 10^{-12} second, how long would you need to observe the system to find all of the particles in the middle container?

Q.10.3 An inventor sends a patent application to the Patent office. In it (s)he claims to have found a chemical compound ensures eternal youth for all who drink it. The patent office rejects it on the basis that the formulation is a perpetual motion machine. Explain why. Did they reject it because it was a perpetual motion machine of the first or second kind?

Q.10.4 An inventor comes to you with a proposal for a new machine that appears to be a perpetual motion machine. Before you throw him out, he begs your indulgence for a moment. Against your better judgment, you relent. He proposes to use the positive volume change of ice on freezing to power a machine. His machine works as follows. A quantity of liquid water is located in a cylinder that has a piston and a cooling unit associated with it. Initially, the cooling unit freezes the water, and as the ice expands, work is done on the piston by raising it. This mechanical work is converted to electrical power by a generator turned by the piston's movement and stored in a battery. As the ice melts, the piston moves downward, the movement again being captured by a differential gear and turning the generator. At the instant the ice becomes liquid, i.e., 273 K, the cooling unit comes on, powered by the battery. As the ice again forms, the piston moves and work is done, and so on. Should you invest in this invention? Why or why not?

This chapter from *The Physical Basis of Biochemistry: Solutions Manual to the Second Edition* corresponds to Chapter 11 from *The Physical Basis of Biochemistry, Second Edition*

P.R. Bergethon, K. Hallock, *The Physical Basis of Biochemistry*,
DOI 10.1007/978-1-4419-7364-1_10, © Springer Science+Business Media, LLC 2011

Q.10.5   How much work is done expanding a 10 cm radius cylinder 50 cm against atmospheric pressure (101.325 kPa)?

Q.10.6   What is the efficiency of a Carnot engine with (a) a cold reservoir at 100 K and a hot reservoir at 400 K? (b) What if the hot reservoir was 1000 K?

Q.10.7   1.2 moles of nitrogen in a 10-cm radius cylinder is expanded at constant temperature from 50 cm to 100 cm against atmospheric pressure. What is the system's change in entropy? (Treat the nitrogen as an ideal gas).

Q.10.8   What do the molecular kinetic and potential energy compare to at the systems level?

Q.10.9   What is the ensemble method?

Q.10.10   What is the ergodic hypothesis?

Q.10.11   In the following canonical ensemble, calculate the total energy, average energy, the total number of elements, and the degeneracy of each energy state? The energy of each ensemble element is listed in its box with units of kJ.

5	1	5	7	2	4	7	5
3	2	6	3	8	6	5	4
1	2	5	7	6	3	3	3
5	4	7	8	4	4	3	8
4	1	3	6	8	3	2	4
6	2	5	8	7	4	4	2
3	7	4	5	1	5	6	5
7	4	6	5	2	6	1	6

Q.10.12   How many different ways could the distinguishable systems shown in Q.10.11 be distributed?

Q.10.13   If one were to select an ensemble element from Q.10.11 at random, what's the probability of selecting each energy level?

Q.10.14   If the size of the ensemble shown in Q.10.11 was increased to 320, but the probabilities of each level remained the same, what would be the total energy, average energy, the total number of elements, and the degeneracy of each energy state? Which of these variables are intensive and which are extensive?

10.2 Answers

A.10.1 (a) $1 : 3^1 = \frac{1}{3}$

(b) $1 : 3^{10} = \frac{1}{59,049}$

(c) $1 : 3^{100} = 1/5.15 \times 10^{47}$

A.10.2 (a) For one particle, it would require 3 picoseconds of observation.

(b) For ten particles, every 59 ns of observation will capture the 1 in 59,049 events.

(c) For 100 particles, it will take a little longer: There are 86,400 s in a day and 31,536,000 s in a year. There are 3.1536×10^{21} picoseconds in a century. Therefore it will take 1.633×10^{26} centuries of looking to find this condition once.

A.10.3 Elixirs of youth are perpetual motion machines of the second kind. They violate the second law of thermodynamics. Life is highly organized and requires energy to maintain this order, a "machine" (the fountain of youth) that can magically create order out of the normal process of the forward arrow of time must violate the second law.

A.10.4 This is a perpetual motion machine of the first kind and violates the first law of thermodynamics. The transfers of energy that occur in this machine will never be able to overcome the frictional losses that are inevitable. The machine will run down. Even if it were possible to make it completely loss-less, it would never be able to produce more energy than was put into it and so at best it would be a curiosity and not a useful or practical machine. Alas even the case of 100% efficiency is impossible.

A.10.5 We must first calculate the volume change; a cylinder's volume is equal to its length multiplied by the area of the circle ($A = \pi \times r^2$). $w = -P_{ext}\Delta V$. $\Delta V = 50$ cm $\times \pi \times (10$ cm$)^2 \times 1$ m$^3/10^6$cm$^3 = +0.0157$ m^3; $dw = -1.01325 \times 10^5$Pa $\times 0.0157$ m$^3 = -1590.8$ J.

A.10.6 (a) Efficiency $= 1 - T_{cold}/T_{hot} = 1 - 100/400 = 0.75$; (b) Efficiency $= 1 - T_{cold}/T_{hot} = 1 - 100/1000 = 0.90$.

A.10.7 Using $\Delta S = n \times R \times \ln(V_2/V_1)$ and knowing that a cylinder's volume is equal to its length multiplied by the area of the circle ($A = \pi \times r^2$), we can calculate the change in entropy. $n = 1.2$ moles. $V_1 = 50$cm $\times \pi \times (10$ cm$)^2 = 15,700$ cm^3; $V_2 = 100$ cm $\times \pi \times (10$ cm$)^2 = 31,400$ cm^3; $\Delta S = n \times R \times \ln(V_2/V_1) = 1.2$ moles $\times 8.314$ J/K $\cdot$ mole $\times \ln(31400/15700) = 6.92$ J/K.

A.10.8 Internal energy and entropy.

A.10.9 The ensemble method is essentially a gedanken (thought) experiment with a large number of imaginary systems. Each system is constructed as a replica on a macroscopic or thermodynamic level of real system whose properties are under investigation.

A.10.10 The ergodic hypothesis states that a single isolated system spends equal
time in each of the available quantum states.

A.10.11

5	1	5	7	2	4	7	5
3	2	6	3	8	6	5	4
1	2	5	7	6	3	3	3
5	4	7	8	4	4	3	8
4	1	3	6	8	3	2	4
6	2	5	8	7	4	4	2
3	7	4	5	1	5	6	5
7	4	6	5	2	6	1	6

The total energy is found by summing the energy of each element. The
average is found by dividing the total by the total number of elementals.
The degeneracy is the total number of elements that are identical, and
in this case, we are only interested in energy so every element with the
same energy is considered degenerate. $E_t = 288$ kJ, $E_{avg} = 4.5$ kJ, $N = 64, n_1 = 5, n_2 = 7, n_3 = 9, n_4 = 11, n_5 = 11, n_6 = 9, n_7 = 7, n_8 = 5$.

A.10.12 Use $W(n) = \dfrac{N!}{\prod_{j=1}^{r} N_j!}$. Because there are 64 elements, we expect the
number of possible distributions to be high. $64!/(5! \times 7! \times 9! \times 11! \times 11! \times 9! \times 7! \times 5!) = 1.65 \times 10^{51}$

A.10.13 To calculate a probability, we divide the degeneracy of an energy level
by the total number of elements. $p_1 = 5/64 = 0.0781, p_2 = 7/64 = 0.1094, p_3 = 9/64 = 0.1406, p_4 = 11/64 = 0.1719, p_5 = 11/64 = 0.1719, p_6 = 9/64 = 0.1406, p_7 = 7/64 = 0.1094, p_8 = 5/64 = 0.0781$.

A.10.14 With the exception of the average, each variable is multiplied by $320/64 = 5$. $E_t = 1440$ kJ, $E_{avg} = 4.5$ kJ, $N = 320, n_1 = 25, n_2 = 35, n_3 = 45, n_4 = 55, n_5 = 55, n_6 = 45, n_7 = 35, n_8 = 25$. With the exception of
average energy, the variables are extensive.

Chapter 11
Which Way Did That System Go?
The Gibbs Free Energy

11.1 Questions

Q.11.1 What is the ΔG of a system at equilibrium?

Q.11.2 A certain ideal gas has a ΔG^0 of -394.65 kJ/mole. What is the ΔG of 1 mole of gas at 10 atm of pressure if temperature remains constant at 20°C?

Q.11.3 What is the chemical potential (μ)? Is it an intensive or extensive variable?

Q.11.4 At 300 K, the ΔG_{mix} of 0.3 moles of ethanol and 0.7 moles of water is -1524 J; determine ΔS_{mix}. (Assume ethanol and water are ideal solutions for this problem.)

Q.11.5 Carbon dioxide is formed by the following chemical reaction:

$$CO + {}^1\!/_2 O_2 \rightleftharpoons CO_2$$

The above reaction has a ΔG^0 of -257.115 kJ/mole at 25°C and 101.3 kPa. What is the equilibrium constant of this reaction?

Q.11.6 Using the value of ΔG^0 provided in Q.11.5, calculate the ΔG of the reaction of 2.3 moles of carbon monoxide with 3 moles of oxygen. Is this reaction spontaneous?

Q.11.7 At some point in the reaction described in Q.11.6, there are 1.9 moles of carbon monoxide, 2.8 moles of oxygen, and 0.4 moles of carbon dioxide. Assuming they are ideal gases, calculate Q.

Q.11.8 How would Q change if the activity coefficient of each gas in Q.11.7 was 0.9?

Q.11.9 If somebody tripled the concentration of oxygen at the time point described in Q.11.7, assuming the gases are ideal, how would the Q change?

This chapter from *The Physical Basis of Biochemistry: Solutions Manual to the Second Edition* corresponds to Chapter 12 from *The Physical Basis of Biochemistry, Second Edition*

P.R. Bergethon, K. Hallock, *The Physical Basis of Biochemistry*,
DOI 10.1007/978-1-4419-7364-1_11, © Springer Science+Business Media, LLC 2011

Q.11.10   The ΔH^0 for the reaction shown in Q.11.5 is -282.989 kJ/mole. What is the change in entropy at 25°C and 101.3 kPa?

11.2 Answers

A.11.1   At equilibrium, there are no changes in energy, heat, and entropy $\Delta G = 0$.

A.11.2   Use $G = G^o + nRT \ln \dfrac{P_2}{P_1}$.

$\Delta G = \Delta G^o + nRT \ln(P_2/P_1) = -394.65$ kJ/mole $+ 1$ mole $\times 8.314$ J/K·mole $\times 293$K $\times \ln(10\text{atm}/1\text{atm})$ 1kJ/1000J $= -389.04$ kJ.

A.11.3   It's the molar Gibb's free energy and it's intensive. Conceptually, chemical potential is to Gibb's free energy what specific heat capacity is to heat capacity; in the former it's per mole and the latter is typically per mass unit.

A.11.4   Using $\Delta G_{mix} = T \times \Delta S_{mix}$ and rearranging,

$$\Delta G_{mix}/T = \Delta S_{mix} = -1524 \text{ J}/300 \text{ K} = -5.08 \text{ J/K}.$$

A.11.5   Use $K = e^{-\Delta G/RT} = e^{-(-257,115 \text{ J/mole}/(298 \text{ K} \times 8.314 \text{ J/K·m})} = 1.18 \times 10^{45}$

A.11.6   Carbon monoxide is the limiting reagent because 1 mole of carbon monoxide reacts with 0.5 moles of O_2. $\Delta G = 2.3$ moles $\times -257.115$ kJ/mole $= -591.4$ kJ. Yes, the reaction is spontaneous because the ΔG is negative.

A.11.7   Use $Q = \dfrac{(a_C)^c (a_d)^d}{(a_A)^a (a_B)^b}$. Because they are ideal gases, the activity coefficient for all of the gases equals 1. Mole fractions are calculated by dividing the moles of one species from the moles of all the species, e.g. X(CO) $= 1.9/(1.9 + 2.8 + 0.4) = 0.37$. X(CO) $= 0.37; \text{X}(O_2) = 0.55; \text{X}(CO_2) = 0.078; 0.078/(0.37 \times 0.55^{1/2}) = 0.284$

A.11.8   Use $Q = \dfrac{(a_C)^c (a_d)^d}{(a_A)^a (a_B)^b}$. Mole fractions are calculated by dividing the moles of one species from the moles of all the species, e.g. X(CO) $= 1.9/(1.9 + 2.8 + 0.4) = 0.37$. X(CO) $= 0.37; \text{X}(O_2) = 0.55; \text{X}(CO_2) = 0.078$; Activity coefficients are used as shown in Equation 12.67. Q $= 0.9 \times 0.37/(0.9 \times 0.078 \times 0.9^{1/2} \times 0.55^{1/2}) = 0.299$

A.11.9   Use $Q = \dfrac{(a_C)^c (a_d)^d}{(a_A)^a (a_B)^b}$. Because they are ideal gases, the activity coefficient for all of the gases equals 1. Moles of oxygen $= 2.8 \times 3 = 8.4$. All of the mole fractions change due to the additional oxygen. Mole fractions are calculated by dividing the moles of one species from the moles of all the species, e.g. $X(CO) = 1.9/(1.9 + 8.4 + 0.4) = 0.18$. $X(CO) = 0.18$; $X(O_2) = 0.79$; $X(CO_2) = 0.037$; $0.037/(0.18 \times 0.79^{1/2}) = 0.238$

A.11.10   Use $\Delta G^o = \Delta H^o - T\Delta S$ and rearrange:

$$\Delta S = (\Delta G^o - \Delta H^o)/T. = (-257.115 \text{ kJ/mole}$$

$$-(-282.989 \text{ kJ/mole})/298\text{K} = 86.8 \text{ J/mol} \cdot \text{K}.$$

Chapter 12
The Thermodynamics of Phase Equilibria

12.1 Questions

Q.12.1 How many degrees of freedom are present in an enclosed system of boiling water?

Q.12.2 Consider the following phase diagram:

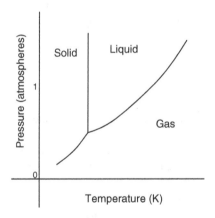

(a) How many points on the diagram do all three phases co-exist?
(b) What is the impact of pressure on the melting point?
(c) What is the impact of pressure on the boiling point?

Q.12.3 The heat of fusion of silver is 104.65 J/g. At 961°C, what is the entropy of fusion of 4.5 g of silver?

Q.12.4 An ideal liquid has a ΔH_{vap} that is independent of temperature. Its boiling point as a function of temperature was determined to be 100°C and 101°C at 101 kPA and 105 kPa, respectively. What is its ΔH_{vap}?

This chapter from *The Physical Basis of Biochemistry: Solutions Manual to the Second Edition* corresponds to Chapter 13 from *The Physical Basis of Biochemistry, Second Edition*

Q.12.5 Urea is considered an eco-friendly de-icer. How many grams of urea (MM = 60.06 g/mol) is required to prevent 3.3 kg of water from freezing at -4°C?

Q.12.6 At 1 atm pressure, how much does a tablespoon of table salt (18 g of NaCl) elevate the boiling point of a pot of water (1.1 l)?

Q.12.7 Assuming it behaves as an ideal solution, what is the osmotic pressure of a 1 μM solution of hemoglobin at 37°C?

Q.12.8 Water striders are insects that skate along the surface of ponds, streams, and other bodies of water. While eating a picnic next to a stream, you watch as somebody dumps a large amount of white powder upstream, as the cloudy water passes underneath the water striders, you notice the insects have trouble staying above the surface. Hypothesize what you think the white powder might be.

Q.12.9 A certain protein has a single binding site with an association constant of 1000 under specific conditions (pH, temperature, etc.). At what concentration of ligand will the binding sites be 25% full? How about 75% full?

Q.12.10 A certain protein has four binding sites, one with an association constant of 1000 and the other three with an association constant of 100. What is the average number of filled sites at 1 mM concentration of ligand? Which type of site contains a higher average number of ligands?

12.2 Answers

A.12.1 Boiling water has two phases: liquid and steam, with water as the sole component. $F = C - P + 2 = 1 - 2 + 2 = 1$.

A.12.2 (a) A single point, known as the triple-point. (b) Increased pressure does not change the melting point because the solid-liquid line is vertical. (c) Increased pressure increases the boiling point.

A.12.3 $T(K) = 961 + 273 = 1234$ K. $\Delta S = \Delta H / T = 104.65$ J/g $\times (1/1234) = 0.0848$ J/g $\cdot$ K. 4.5 g $\times$ 0.0848 J/g $\cdot$ K = 0.3816 J/K. $\Delta S = 0.3816$ J/K

A.12.4 Rearranging the Clausius-Clapeyron equation, we have: $\Delta H = R \times \ln(P_2/P_1)/(1/T_2 - 1/T_1) = 8.314$ J/K $\cdot$ mole $\times \ln(105 \text{ kPa}/101 \text{ kPA})/(1/373$ K $- 1/374$ K$) = 47.05$ kJ.

A.12.5 $m = \Delta T_f / k_f$. For water, $k_f = 1.86$ K/molal. Since the normal freezing point of water is 0°C, $\Delta T_f = 4^\circ$C. m $= 4^\circ$C$/18.86$ K/molal $= 2.15$ molal. Molal = moles of solute/kg of solvent, since we have 3.3 kg of water, we need 2.15 molal $\times$ 3.3 kg $= 7.19$ moles. 60.06 g/mole $\times$ 7.19 moles $= 431.98$ g of urea.

A.12.6 $m \times k_b = \Delta T_b$. $k_b = 0.51$ for water. 1.1 l of water has a 1.1 kg mass. NaCl MM $= 58.44$ g/mole. 18 g/58.44 g/mole $= 0.307$ moles. $m = 0.307$ moles/1.1 kg $= 0.280$ molal. $0.280 \times 0.51 = 0.143^\circ$C

A.12.7 $\pi = cRT = 1$ μM $\times 8.314$ L $\cdot$ kPa/m $\cdot$ K $\times 310$K $= 2.58$ Pa.

A.12.8 One hypothesis is that the white powder is a surface active detergent. The detergent would lower the surface tension making it difficult for the water striders to stay on top of the water.

A.12.9 $v = K_{assoc} \times [A]/(1 + K_{assoc} \times [A])$. Rearranging, $[A] = v/(K_{assoc} + v \times K_{assoc})$. For 25%, $[A] = 333 \ \mu M$. For 75%, $[A] = 3$ mM.

A.12.10 $v = n_1 \times k_1 \times [A]/(1 + k_1 \times [A]) + n_2 \times k_2 \times [A]/(1 + k_2 \times [A])$. $n_1 = 1, k_1 = 1000, n_2 = 3, k_2 = 100, [A] = 0.001. v = 0.773$. The type 1 site contains on average 0.5 ligands, while the three type 2 sites only average 0.273 ligands for all of them, so the type 1 sites contain a higher average of ligands.

Part III
Building a Model of Biomolecular Structure

Chapter 13
Water: A Unique Structure, a Unique Solvent

13.1 Thought Exercises

Q.13.1 From your knowledge of water structure, explain why ice floats. Would you expect ice and water to have the same structure in Boston as at the bottom of the ocean? How would these structures differ? What implications does this have for the biochemicals that might make up a deep-water fish?

Q.13.2 In 1997, great excitement surrounded the presumed existence of water on one of the moons of Jupiter. This moon was thought to have enough heat generated from its volcanic activity that water would be melted underneath a mantle of ice. What is the likely heat capacity, dielectric constant, and heats of formation and vaporization for this Jovian water when compared to terrestrial water?

Q.13.3 A Jovian enzyme is discovered by a space probe that appears to act by adding a water molecule to an ester bond (hydrolysis). Assuming that the enzyme maintains its structure when brought back to Earth, comment on the relative activity of the hydrolysing water in the terrestrial environment.

Q.13.4 The concept of the cell as a container of dilute aqueous solution can be supported only if certain abstractions are made. List some of these.

This chapter from *The Physical Basis of Biochemistry: Solutions Manual to the Second Edition* corresponds to Chapter 14 from *The Physical Basis of Biochemistry, Second Edition*

P.R. Bergethon, K. Hallock, *The Physical Basis of Biochemistry*, 63
DOI 10.1007/978-1-4419-7364-1_13, © Springer Science+Business Media, LLC 2011

Chapter 14
Ion-Solvent Interactions

14.1 Questions

Q.14.1 List three assumptions of the Born model.

Q.14.2 Describe some of the atomic properties of water. Is it a structureless continuum?

Q.14.3 Using the Born approximation, calculate the ΔG_{i-s} of 0.6 moles of Ca^{2+} in water; its ionic radius is 114 pm.

Q.14.4 Using the Born approximation, calculate the ΔH_{i-s} of 0.3 moles of Na^+ in water at 283 K and 353 K; its ionic radius is 186 pm. The dielectric of water changes with temperature; ε (283 K) $= 84.11$ and ε (353 K) $= 63.80$.

Q.14.5 The answers from Q.14.3 and Q.14.4 are substantially different than the experimentally measured quantities, why? What needs to be included to make the calculated values more accurate?

Q.14.6 Why is the heat capacity of a 1 molar solution of KBr substantially lower than a 0.5 molar solution of KBr?

14.2 Thought Exercise

Name two situations in biological systems where ion pairing as described by Bjerrum may be of consequence. Justify your answer.

14.3 Thought Exercise

Lithium chloride and urea are used in the isolation of RNA. Based on your knowledge of aqueous solutions, why have these substances been found so effective?

This chapter from *The Physical Basis of Biochemistry: Solutions Manual to the Second Edition* corresponds to Chapter 15 from *The Physical Basis of Biochemistry, Second Edition*

P.R. Bergethon, K. Hallock, *The Physical Basis of Biochemistry*,
DOI 10.1007/978-1-4419-7364-1_14, © Springer Science+Business Media, LLC 2011

14.4 Answers

A.14.1   (1) The ion may be represented as a rigid sphere of radius r_i and charge ze_0, where z_i is the charge number and e_0 is the charge on an electron. (2) The solvent into which the ion is to be dissolved can be treated as a structureless continuum. (3) All interactions are electrostatic in nature.

A.14.2   Water is not a structureless continuum. Water is a polar molecule with a dipole moment of 1.85 debye. Water is ideal for forming hydrogen bonds, and in bulk water these bonds are extensively formed with other water molecules

A.14.3   Use Equation 15.10 and Table 14.2. $z = 2$ because the calcium ion has a +2 charge. $e = 1.60 \times 10^{-19}$ C, $z = 2$, $\varepsilon_0 = 8.85 \times 10^{-12}$ F/m, $\varepsilon = 78.5$, $\pi = 3.14$, $N_a = 6.022 \times 10^{23}$, number of moles $= 0.6$ moles, radius $= 114$ pm; $\Delta G = -1.44 \times 10^6$ J

A.14.4   Use Equation 15.10 and Table 14.2. $z = 1$ because the sodium ion has a +1 charge. Because water's dielectric constant changes with temperature, we must include the rate of that change with temperature ($d\epsilon/dT$) as part of our calculation. This can be estimated by assuming the change is linear with temperature over the temperature range of the problem and estimating $d\epsilon/dT$ from $\Delta\varepsilon/T = (63.80 - 84.11)/70$ K $= -0.29/$ K. $e = 1.60 \times 10^{-19}$ C. $z = 1$, $\varepsilon_0 = 8.85 \times 10^{-12}$ F/m, $\varepsilon = 84.11$ or 63.80 (at 283 K and 353 K, respectively), $\pi = 3.14$, $N_a = 6.022 \times 10^{23}$, number of moles $= 0.3$ moles, radius $= 186$ pm, T (K) $= 280$ or 350, $\Delta\varepsilon/T = -0.29/$K, $\Delta H(280) = -111.8 \times 10^3$ J, $\Delta H(350) = -112.8 \times 10^3$ J.

A.14.5   The Born approximation assumes water is a continuum and that the rigid spheres with a radius equal to the crystallographic radius, but neither of these is correct when an ion is dissolved in water. Water reorganizes around the ion adding a hydration sheath and increasing its effective radius. Including these factors, as well as some others, would make the theory more accurate.

A.14.6   The presence of KBr reduces the number of hydrogen bonds within the solution. Hydrogen bonds can absorb a lot of thermal energy, so when they're removed, the amount of heat that can be absorbed by the solution is decreased. The greater the amount of KBr, the lower the number of hydrogen bonds.

Chapter 15
Ion-Ion Interactions

15.1 Questions

Q.15.1 What is the ionic strength of the serum component of blood? What is the osmolarity of this component? Normal Serum composition is: Na^+ 140 milliequivalents, K^+ 4 milliequivalents, Cl^- 100 milliequivalents, HCO_3^- 24 milliequivalents.

Q.15.2 Calculate the effective radius of the ionic atmosphere for ions in the serum. Base your answer on the work derived in Question Q.15.1

Q.15.3 What is the ionic strength of the following?

(a) 0.5 M KCl
(b) 0.5 M $CaCl_2$
(c) 0.1 M K_2MnO_4
(d) 0.3 M H_2CO_3
(e) 0.01 M $Ba(OH)_2$

Q.15.4 What is the radius of the Debye atmosphere for the cations in question 4 at 298 K? at 310 K? at 400 K? at 274 K?

Q.15.5 List three assumptions of the Debye-Hückel model.

Q.15.6 What is the principle of electroneutrality and how does it relate to ions dissolved in solution?

Q.15.7 What assumption of the Debye-Hückel model permits the use of the linearized Boltzmann equation? Is this assumption valid if $z_i e_o \psi_r / kT = 0.5$? 0.1? 0.01?

Q.15.8 Why does the Debye length increase with temperature?

Q.15.9 Is the Debye-Hückel model a good model for ionic interactions within a biological system?

This chapter from *The Physical Basis of Biochemistry: Solutions Manual to the Second Edition* corresponds to Chapter 16 from *The Physical Basis of Biochemistry, Second Edition*

P.R. Bergethon, K. Hallock, *The Physical Basis of Biochemistry*, DOI 10.1007/978-1-4419-7364-1_15, © Springer Science+Business Media, LLC 2011

15.2 Answers

A.15.1 Although ionic strength can be calculated using any concentrations, molarity is a more convenient unit so the first step is to convert each milliequivalents concentration into molarity. For monovalent ions, 1 milliequivalent = 1 millimolar. Then sum up the terms needed to calculate the ionic strength:

$$I = \frac{1}{2} \sum c_i z_i^2$$

$$I = 0.5 \times (140 \text{ mM} \times (+1)^2 + 4 \text{ mM} \times (+1)^2 + 100 \text{ mM} \times (-1)^2$$

$$+ 24 \text{ mM} \times (-1)^2) = 134 \text{ mM}$$

Assuming the ions behave ideally, osmolarity can be calculated by summing up the concentration of particles in solution: osmolarity = 140 mM + 4 mM + 100 mM + 24 mM = 268 mM.

A.15.2 See Q.15.4 for a detailed explanation of the unit analysis required for calculating Debye lengths.

The Debye length varies based on ionic strength and temperature as shown in this formula. We'll assume the serum is at 298 K, which means the dielectric constant of water is 78.5. (Table 14.2)

$$\kappa^{-1} = L_D = \left(\frac{\varepsilon \varepsilon_0 kT}{2 e_o^2 N_A I} \right)^{1/2}$$

$\varepsilon = 78.5$ (unitless)
$\varepsilon_0 = 8.854 \times 10^{-12} \text{J}^{-1} \text{C}^2 \text{m}^{-1}$
$k = 1.381 \times 10^{-23} \text{J K}^{-1}$
$T = 298 \text{ K}$
$e_0 = 1.602 \times 10^{-19} \text{C}$
$N_A = 6.022 \times 10^{23} \text{mole}^{-1}$
$I = 134 \text{ mM} = 0.134 \text{ mole dm}^{-3}$

Using the above values to calculate the Debye length, we find $L_D = 0.83$ nm.

A.15.3 $I = \frac{1}{2} \sum c_i z_i^2$ (a) $0.5 \times (0.5 \times (+1)^2 + 0.5 \times (-1)^2) = 0.5 \text{ M}$. (b) $0.5 \times (0.5 \times (+2)^2 + 1 \times (-1)^2) = 1.5 \text{ M}$, (c) 0.3 M, (d) 0.9 M, (e) 0.03 M.

A.15.4 The Debye length varies based on ionic strength and temperature as shown in this formula. It also depends on the dielectric constant of water, which varies some with temperature, but we'll ignore that variation and use 78.5 for all the temperatures.

$$\kappa^{-1} = L_D = \left(\frac{\varepsilon \varepsilon_0 kT}{2 e_o^2 N_A I} \right)^{1/2}$$

We will digress just a little to mention the importance of unit analysis when calculating the Debye length. Here are all of the terms and their respective units.

$$\varepsilon = 78.5 \text{ (unitless)}$$
$$\varepsilon_0 = 8.854 \times 10^{-12} J^{-1} C^2 m^{-1}$$
$$k = 1.381 \times 10^{-23} J\, K^{-1}$$
$$T = 298\, K$$
$$e_0 = 1.602 \times 10^{-19} C$$
$$N_A = 6.022 \times 10^{23} mole^{-1}$$
$$I = 1 \text{ mole}/1 = 1 \text{ mole dm}^{-3}$$

T and I were set at arbitrary values for illustrative purposes. If the unit analysis is done incorrectly, which can happen if I is used with the units of molarity and is not converted to m^3, the result number will be incorrect by a few orders of magnitude.

$$((78.5 \times 8.854 \times 10^{-12} J^{-1} C^2 m^{-1} \times 1.381 \times 10^{-23} J\, K^{-1} \times 300\, K)/$$
$$((1.602 \times 10^{-19} C)^2 \times 6.022 \times 10^{23} mole^{-1} \times 1 \text{ mole}/dm^3))^{1/2}$$

To simplify the equation, we will collect all of the numbers into a Constant and treat only the units. You have to convert molarity, which is moles/dm^3, into a m^3, which is done using the relationship $1\, m = 10\, dm$.

$$= (\text{Constant} \times (J^{-1} C^2 m^{-1} \times J\, K^{-1} \times K)/(C^2 \times mole^{-1} \times mole\, dm^{-3}))^{1/2}$$
$$= (\text{Constant} \times m^{-1} \times dm^3)^{1/2}$$
$$= (\text{Constant} \times m^{-1} \times dm^3 \times 1 m^3 \times (10 dm)^{-3})^{1/2}$$
$$= (\text{Constant} \times m^2)^{1/2}$$
$$= (\text{Constant})^{1/2} m$$

The final unit is meters, which is the desired unit of distance. To simplify the unit analysis, some books change I into moles/m^3, which is mathematically equivalent, but in this book, we report I in molarity and convert the units as part of the calculation. The following are the answers for each of the problems using the above technique.

At 298 K (a) 0.43 nm, (b) 0.25 nm, (c) 0.56 nm, (d) 0.32 nm, (e) 1.76 nm.

At 310 K (a) 0.44 nm, (b) 0.25 nm, (c) 0.57 nm, (d) 0.33 nm, (e) 1.79 nm.

At 400 K (a) 0.50 nm, (b) 0.29 nm, (c) 0.64 nm, (d) 0.37 nm, (e) 2.03 nm.

At 274 K (a) 0.41 nm, (b) 0.24 nm, (c) 0.53 nm, (d) 0.31 nm, (e) 1.68 nm.

A.15.5 Three of the following. (1) A central reference ion of a specific charge can be represented as a point charge. (2) This central ion is surrounded by a cloud of smeared-out charge contributed by the participation of all of the other ions in solution. (3) The electrostatic potential field in the solution can be described by an equation that combines and linearizes the Poisson and Boltzmann equations. (4) No ion − ion interactions except the electrostatic interaction given by a $1/r^2$ dependence are to be considered (i.e., dispersion forces and ion − dipole forces are to be excluded). (5) The solvent simply provides a dielectric medium, and the ion − solvent interactions are to be ignored, so that the bulk permittivity of the solvent can be used.

A.15.6 A solution of ions will be electroneutral because each of the central ions will be surrounded by atmospheres of charge that are exactly equal in magnitude but opposite in sign to the charge on the central ion. Mathematically, the principle can be expressed as: $\sum z_i e_0 X_i = 0$.

A.15.7 The linearized Boltzmann equation can be used if $z_i e_o \psi_r / kT$ is much smaller than 1. This is because the third term in the Taylor series expansion squares $z_i e_o \psi_r / kT$, making the term negligible if it is very small. If $z_i e_o \psi_r / kT = 0.1$, $\rho_r = 1 - 0.1 - 0.005$. The third term isn't negligible so the assumption wouldn't be valid. If $z_i e_o \psi_r / kT = 0.01$, $\rho_r = 1 - 0.01 - 0.00005$. The third term is small relative to the first two so the assumption might be reasonable. If $z_i e_o \psi_r / kT = 0.01$, $\rho_r = 1 - 0.01 - 0.00005$. The third term is very small when compared to the first two so the assumption would be reasonable.

A.15.8 $\kappa^{-1} = L_D = \left(\dfrac{\varepsilon \varepsilon_o kT}{2 e_o^2 N_A I} \right)^{1/2}$

The Debye length increases as the square root of temperature. Because the Debye length represents the effective volume occupied by an ion, it would be expected to increase with temperature because the average kinetic energy of an ion increases with temperature. At higher temperatures, ions are moving faster, effectively occupying larger volumes than they do at lower temperatures, increasing their effective radius, i.e. their Debye length.

A.15.9 The Debye-Hückel limiting law suggests it is not a good model because the ionic strength of solutions in vivo are too high to be accurately modeled by Debye-Hückel model.

Chapter 16
Lipids in Aqueous Solution

16.1 Questions

Q.16.1 Define the critical micelle concentration

Q.16.2 Draw (a) a lipid micelle, (b) a phospholipid bilayer.

Q.16.3 Study Fig. 17.9, what are the major differences between the phosphatidic acid (PA), phosphatidylcholine (PC), and phosphatidylethanolamine (PE) headgroups?

Q.16.4 Comparing PA, PC, and PE (Fig. 17.9), which has the smallest headgroup and which has the largest? Why?

Q.16.5 A scientist creates some phospholipid liposomes (Fig. 17.3c) by mixing PE and PC; which phospholipid will most likely occupy the inside layer?

Q.16.6 Assuming everything else between two fatty acids is identical, would you expect a fatty acid with (a) 0 or 3 double bonds to have a lower melting point? (b) one with a 10 carbon chain or 20 carbon chain?

Q.16.7 List two types of diffusion lipid molecules can undergo in a liquid crystalline phase.

Q.16.8 Why is transmembrane flip-flop of a lipid molecule in a bilayer energetically unfavorable?

Q.16.9 Study Fig. 17.11. Based on its structure, do you think cholesterol will more closely associate with the polar headgroup or non-polar acyl chain regions of a lipid bilayer?

Q.16.10 What is the fluid-mosaic model?

16.2 Thought Exercise

Propose the key elements of a transmembrane protein that transports water through a lipid bilayer (consider the aquaporins). Draw a picture.

This chapter from *The Physical Basis of Biochemistry: Solutions Manual to the Second Edition* corresponds to Chapter 17 from *The Physical Basis of Biochemistry, Second Edition*

P.R. Bergethon, K. Hallock, *The Physical Basis of Biochemistry*,
DOI 10.1007/978-1-4419-7364-1_16, © Springer Science+Business Media, LLC 2011

16.3 Answers

A.16.1   It's the concentration at which large organic ions change from solubilized monomeric form to micelles when dissolved in water.

A.16.2   See Figs. 17.4 and 17.13, respectively.

A.16.3   The major difference is the functional group attached to the oxygen on the phosphorous atom. PA has an H attached to it. PE has two carbons and a primary amine, while PC has two carbons and a tertiary amine with three methyl groups. This changes the relative volume of the headgroups which has a substantial impact on the physical properties of the lipids.

A.16.4   Everything between the three is identical except for their functional group. PA has the smallest functional group (-H) and PC has the largest ($-N(CH_3)_3$), so PA is the smallest and PC is the largest.

A.16.5   The space for headgroups is most limited on the inside liposome layer so PE, with its smaller headgroup, will most likely occupy the inside layer.

A.16.6   Based on Fig. 16.7, fatty acids with (a) 3 double bonds and (b) 10 carbon chains will have lower melting points. A likely reason is that packing of the acyl chains will be less organized for the acyl chains with more double bonds and longer carbon chains.

A.16.7   Rotational and lateral diffusion.

A.16.8   There is a large energy barrier to passing a polar headgroup through the non-polar interior of the lipid bilayer.

A.16.9   Cholesterol is mostly non-polar so it will associate more closely with the non-polar region of a lipid bilayer.

A.16.10 The fluid-mosaic model is the modern model of the lipid bilayer in which the various proteins are "floating" in the bilayer structure. Some proteins span the lipid bilayer, while others interact with one side of the membrane or the other.

Chapter 17
Macromolecules in Solution

17.1 Questions

Q.17.1 Consider the concept that primary structure leads to a functional protein as described by Anfinsen's experiments with ribonuclease. Why was it fortunate that Anfinsen worked with ribonuclease? Would the same conclusions been found if the protein insulin had been used.

Q.17.2 What is the ΔG_t from ethanol to water of a peptide comprised of three serines compared to one with three glycines? Instead of serines, calculate it for three phenylalanines? Which peptide is more hydrophobic and why?

Q.17.3 What amino acid sequence do the following nucleic acids encode for? (a) AGA-GUU-GGG, (b) CUU-GAU-AAU, (c) GAA-UAU-CUG , (d) AGG-GUC-GGA.

Q.17.4 The fraction of folded protein for an unknown protein was found to be 0.75 and 0.2 at $T = 25°C$ and $35°C$, respectively. What is the van't Hoff enthalpy?

Q.17.5 What are two limiting-case models for protein folding? How do they differ?

Q.17.6 What is the primary component of the amyloid plaques associated with Alzheimer's disease?

17.2 Answers

A.17.1 The unfolding and refolding of ribonuclease required a protein to be able to search its various "disulfide" conformation states with the assumption that the backbone of the protein always returned to a "ground" state from which to restart searching the folding space. Insulin as a protein, folds with a number of disulfide links stabilizing it and then undergoes a cleavage of

This chapter from *The Physical Basis of Biochemistry: Solutions Manual to the Second Edition* corresponds to Chapter 18 from *The Physical Basis of Biochemistry, Second Edition*

P.R. Bergethon, K. Hallock, *The Physical Basis of Biochemistry*,
DOI 10.1007/978-1-4419-7364-1_17, © Springer Science+Business Media, LLC 2011

the folded protein. If insulin is denatured via reduction, the result is two chains when there had been one. If Anfinsen had done his experiment with insulin, he would not have been able to retrieve the correct statistical values that were essential to proving his case.

A.17.2 According to Table 18.6, each amino acid has a different ΔG of transfer and glycine is set to zero for convenience. Multiplication is all that's required to approximate the answers to these questions, assuming that the three amino acids have similar thermodynamic properties to their individual components. Serines: $3 \times (-1.3 \text{ J mol}^{-1}\text{deg}^{-1}) = -3.9 \text{ J mol}^{-1} \text{deg}^{-1}$, Phenylalanines $= 3 \times (10.5 \text{ J mol}^{-1}\text{deg}^{-1}) = 31.5 \text{ J mol}^{-1} \text{deg}^{-1}$ The phenylalanine peptide is more hydrophobic because its ΔG_t from ethanol to water more positive, suggesting it energetically prefers to be in ethanol compared to water.

A.17.3 Comparing each three nucleic acid code sequence with Table 18.3 will translate the nucleic acids into amino acids. (a) Arg-Val-Gly, (b) Leu-Asp-Asn, (c) Glu-Tyr-Leu, (d) Arg-Val-Gly.

A.17.4 Equation (18.19) can be used to calculate the van't Hoff enthalpy. The key is to plot $\ln K$ vs. $1/T$, with the slope giving $-\Delta H/R$. The two sets of data presented in the problem can be used to create a two-point line. $T1 = 298.15 \text{ K}$, $T2 = 308.15 \text{ K}$, $1/T1 = 3.35 \times 10^{-3}$, $1/T2 = 3.25 \times 10^{-3}$; $K1 = 0.75/(1 - 0.75) = 3$, $K2 = 0.2/(1 - 0.2) = 0.25$; $\ln K1 = 1.10$, $\ln K2 = -1.39$. Find the slope using a calculator or computer program like Excel. Slope $= 22,830 \text{ K}^{-1}$. $-\Delta H/R = 22,830 \text{ K}^{-1}$, $\Delta H = -189.8$ kJ/mole.

A.17.5 The sequential model and the nucleation growth model. The sequential model proposes a step-by-step model in which there is linear movement within structural state space from unfolded to folded. Models like the nucleation growth model propose discontinuous folding during which a protein needs to create a critical nucleation point before collapsing into its folding confirmation.

A.17.6 The amyloid β protein that has misfolded to form fibrils.

Chapter 18
Molecular Modeling – Mapping Biochemical State Space

18.1 Questions

Q.18.1　Is further modification of the Ramachandran plot dependent on the overall volume of the R group? Why or why not?

Q.18.2　Propose an atom type classification for oxygen that would be useful in biochemical investigations.

Q.18.3　Define the following terms: (a) United force field, (b) All atom force field

Q.18.4　For theoretical calculations, limits on computational time forces researchers to decide which systems need all atom force fields to be accurately modeled and which ones can be modeled using united force fields. Assuming you are a researcher who only has enough computational power to use all atom force fields on three of the six following systems, which of them would be most accurately modeled using united force fields and why?

Phospholipids located in a membrane, phospholipids (as second messengers) in the cytosol, sphingomyelin, folded protein structures, reactive sites of enzymes, nucleic acids in the zinc finger of the chromosome.

Q.18.5　What are the assumptions and constraints for each of these molecular modeling systems? (a) Force field methods, (b) Semi-empirical method, (c) Ab initio method.

Q.18.6　Draw the lowest and highest energy configurations of n-butane. Using steric hindrance, explain your choices.

Q.18.7　Estimate a Ramachandran plot for a part of a protein that has ten α-helix amino acids and ten β-sheet amino acids.

Q.18.8　A researcher determined the following amino acid sequence that is part of a water soluble protein: GIGAELKVLKKGGPAEISWIKRKRQQ. Using Tables 19.4 & 19.5, predict the secondary structure of this stretch of the protein at STP when it is dissolved in water. How might you determine the accuracy of your prediction?

This chapter from *The Physical Basis of Biochemistry: Solutions Manual to the Second Edition* corresponds to Chapter 19 from *The Physical Basis of Biochemistry, Second Edition*

P.R. Bergethon, K. Hallock, *The Physical Basis of Biochemistry*,
DOI 10.1007/978-1-4419-7364-1_18, © Springer Science+Business Media, LLC 2011

Q.18.9 O_2 has a equilibrium bond length of 1.208 Å at STP. (CRC Handbook of Chemistry and Physics 71st edition, 9–2) with $K_r = 1177\,N\,m^{-1}$. (Symmetry and Spectroscopy, *An Introduction to Vibrational and Electronic Spectroscopy*, Harris and Bertolucci, p. 105). Using Hooke's Law, calculate the potential energy change when the bond distance is increased by (A) 0.001 Å, (B) 0.005 Å. What frequency photon would stretch the bond to (A) 0.001 Å, (B) 0.005 Å. Is the O_2 bond stronger or weaker than the N_2 bond (Nitrogen's $K_r = 2297\,N\,m^{-1}$)?

Q.18.10   What is the primary difference between the Extended Hückel and Hückel semi-empirical methods? Why would somebody use a simpler approximation instead of one that is more accurate?

18.2 Answers

A.18.1   No. The volume of the R groups is a short-range interaction that is already considered when making Ramachandran plots. Improvements are better made by considering longer range interactions such as electrostatic interactions.

A.18.2   There several different ways to do this; we will use the number of different carbons types as our guide and limit ourselves to four types of oxygen. One possible classification would have alcohols (-COH), carbonyls (-COOH), ethers (-COC-) and esters (-CO).

A.18.3   (a) United force fields treat a collection of atoms as an atom type, treating a group of atoms and one single entity, e.g. a methyl group would be treated as one entity instead of three hydrogens and a carbon. (b) All atom force fields treat every atom individually.

A.18.4   The three systems that would be the best choices to model using united force fields are: phospholipids located in a membrane, sphingomyelin, and folded protein structures. Phospholipids and sphingomyelin contain long acyl chains that are not essential to any reactions, so they could be modeled using united force fields. Hydrophobicity is an important energetic constraint on protein folding, but the atomic nature that gives rise to it is less important and could be modeled using united force fields.

Reactive sites of enzymes and nucleic acids in the zinc finger of the chromosome are better modeled if all of the atoms are considered because so many small interactions contribute to the overall behavior of these systems making it more essential to use all atom force fields. Unlike phospholipids in the membrane, phospholipids (as second messengers) in the cytosol may interact specifically with certain proteins and the exact nature of their acyl chain may play a role in those interactions, so it would be better modeled using that all atom force fields.

A.18.5   (a) Empirical methods utilize the bonding derived from quantum mechanical calculations, but specify the properties of the bonds linking atoms, and the atomic properties of the objects linked in coordinate space, based on measured physical properties. These techniques are built on experimental methods and provide the mechanical properties of molecules which are used in a Newtonian framework to model molecules of interest. Because they assume the molecule's behavior is based on already known experimental results, they are less likely to identify new behaviors, which limits its usefulness as a theoretical research tool. (b) Semi-empirical methods are constrained by the experimental results to help reduce the computational time, but still uses ab initio calculations to optimize the calculated structure. Like ab initio, it doesn't make any a priori assumption of such things as chemical bonds but rather the bonding behavior is a result of the quantum mechanical calculations. (c) Ab initio techniques are all based on the molecular orbital methods of LCAO-MO. It doesn't make any a priori assumption of such things as chemical bonds but rather the bonding behavior is a result of the quantum mechanical calculations, and doesn't use experimental data to limit molecular behavior. It is the technique that uses the most quantum mechanics and is the most computationally intensive so it is constrained by how rapidly the calculations can be performed.

A.18.6   Refer to Fig. 19.2 in the textbook. The lowest energy is when the methyl groups they are furthest from each other and the highest energy is when they're closest. The methyl groups are relatively bulky so the more room they have, the lower the energy state.

A.18.7

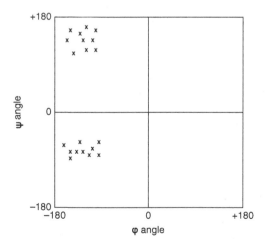

A.18.8   Each amino acid needs to have its parameters calculated and then the guidelines in the tables need to be consulted to identify the most likely secondary structure.

#	Amino acid	P-alpha	6 AA P-alpha avg	P-beta	6 AA P-beta avg	Structure
Six Amino Acid Avg = Central AA + AA(N + 2) + AA (N−1) + AA(N + 2) + AA(N−2) + 0.5*AA(N + 3) + 0.5*AA(N−3)						
1	Gly	0.53	0.46	0.81	0.62	random coil
2	Ile	1.00	0.71	1.60	0.72	random coil
3	Gly	0.53	0.95	0.81	0.84	random coil
4	Ala	1.45	1.11	0.97	0.94	α-helix
5	Glu	1.53	1.17	0.26	0.94	α-helix
6	Leu	1.34	1.24	1.22	0.98	α-helix
7	Lys	1.07	1.28	0.74	0.99	α-helix
8	Val	1.14	1.21	1.65	1.01	α-helix
9	Leu	1.34	1.10	1.22	1.02	α-helix
10	Lys	1.07	0.99	0.74	0.99	random coil
11	Lys	1.07	0.90	0.74	0.91	random coil
12	Glu	0.53	0.86	0.81	0.80	random coil
13	Gly	0.53	0.91	0.81	0.74	random coil
14	Pro	0.59	0.94	0.62	0.77	random coil
15	Ala	1.45	0.96	0.97	0.84	random coil
16	Glu	1.53	1.03	0.26	0.86	α-helix
17	Ile	1.00	1.16	1.60	0.86	α-helix
18	Ser	0.79	1.21	0.72	0.81	α-helix
19	Trp	1.14	1.12	1.19	0.85	α-helix
20	Glu	1.53	1.06	0.26	0.83	α-helix
21	Lys	1.07	1.07	0.74	0.77	α-helix
22	Arg	0.79	1.07	0.90	0.79	α-helix
23	Lys	1.07	1.04	0.74	0.88	α-helix
24	Arg	0.79	0.92	0.90	0.90	random coil
25	Gln	1.17	0.77	1.23	0.76	random coil
26	Gln	1.17	0.61	1.23	0.62	random coil

Possible techniques for measuring the accuracy of the prediction include circular dichroism spectroscopy, nuclear magnetic spectroscopy, and x-ray crystallography.

A.18.9   The potential energy is calculated as if it were a spring, which isn't accurate, but is sufficient for the purposes of this problem.

Use Equation 19.3: $U_{stretch} = \sum K_r(r - r_0)^2$; $E_{photon} = h\nu$;

$1\,J = 1\,N \times m$; $1\,\text{Å} = 10^{-10}\,m$

(a) $1177\,N\text{-}m^{-1} \times (0.001\,\text{Å})^2 = 1177\,J\text{-}m^{-2} \times (1 \times 10^{-13} m)^2 = 1.18 \times 10^{-23}\,J$

Photon Frequency $= 1.78 \times 10^{10}\,Hz$

(b) $1177\,\text{N - m}^{-1} \times (0.005\,\text{Å}) = 1177\,\text{J - m}^{-2} \times (5 \times 10^{-13}\text{m})^2 = 2.94 \times 10^{-22}\text{J}$

Photon Frequency $= 4.44 \times 10^{11}\text{Hz}$

Nitrogen's triple bond is stronger than oxygen's double bond because its force constant is higher.

A.18.10 The Extended Hückel recognizes the σ interactions with π electrons, while the Hückel ignores them. More accurate approximations are computationally more expensive so researchers must decide the best method to answer their research question. Simpler models are often adequate.

Chapter 19
The Electrified Interphase

19.1 Questions

Q.19.1   Assume that an isolated preparation of cytosolic organelles has been obtained and is sitting in a test tube on your bench. Describe and discuss two methods for precipitating these organelles based on their colloidal nature. Discuss when one method might be preferred over the other.

Q.19.2   Describe a molecular model for the organization of water molecules around a cell. Include in your analysis the orientation, structure, dimensions, and composition of the region.

Q.19.3   Is the Stern or Grahame model of the interphase to be preferred over the Gouy − Chapman or Helmholtz model in biological systems?

Q.19.4   In experiments to investigate the protein structure of the nucleosome octamer, the isolated nucleosome is placed in a very high ionic strength solution. This step causes the dissociation of the DNA from the histones. Explain this in terms of the interactional energies of nucleosome formation.

Q.19.5   The voltage change across the plasma membrane in a firing neuron is $-80\,mV\, -->\, +50\,mV$.

    (a)   What is the change in electric field when the neuron fires? The membrane is 6 nanometers thick.

    (b)   The capacitance of the membrane is $1 \times 10^{-6}\,F$. How many charges pass through the membrane to cause the measured voltage change?

Q.19.6   Why is the ΔS of movement to the inner Helmholtz plane positive for Na^+ and K^+ and negative for Cl^- and I^-?

Q.19.7   Describe the Helmholtz-Perrin model and discuss one of its fundamental problems. Draw the arrangement of counter ions and electrode as proposed by the Helmholtz-Perrin model.

This chapter from *The Physical Basis of Biochemistry: Solutions Manual to the Second Edition* corresponds to Chapter 20 from *The Physical Basis of Biochemistry, Second Edition*

P.R. Bergethon, K. Hallock, *The Physical Basis of Biochemistry*,
DOI 10.1007/978-1-4419-7364-1_19, © Springer Science+Business Media, LLC 2011

Q.19.8   Describe the Gouy-Chapman model and discuss one of its fundamental problems. Draw the arrangement of counter ions and electrode as proposed by the Gouy-Chapman model.

Q.19.9   Describe the Stern model and discuss its behavior in dilute ion solutions and concentrated ion solutions. When is it more like the Helmholtz-Perrin model? When is it more like the Gouy-Chapman model? Draw the arrangement of counter ions and electrode as proposed by the Stern model.

Q.19.10 Consider Figs. 20.8 and 20.10, comment on the behavior of the superposition of the interactional force and the Lennard-Jones potential when the interactional force between the double layers ψ_o = (a) 50, (b) 10, (c) 2.5, and (d) 0.5. Sketching graphs may help. (These numbers are on the same arbitrary magnitude scale as the one used in Fig. 20.10.) Propose a mechanism that could change the interactional force between double layers other the infusion of counter ions (which is mentioned in the text.)

19.2 Answers

A.19.1   The precipitation of the colloidal organelles will occur if the ionic atmosphere that surrounds them can be collapsed and made small enough in diameter that the colloids can approach one another and multiple van der Waals interactions can lead to agglutination and precipitation. Raising the ionic strength in the solvent will reduce the Debye atmosphere and increase contact of the colloids hence increasing likelihood of precipitation. A corollary method to reduce the Debye atmosphere is to use higher valance salts such as Ca^{+2} rather than monovalent salts such as sodium and chloride. Yet another method is the reduction of the activity of the solvent (water) by forcing the water to associate with other solutes that are chaotropic agents such as urea and guanidium. These disrupt normal water structure and reduce it activity. The resulting dehydration can enhance the association of the colloidal surfaces.

A.19.2   The water around a cell can be divided into three regions: the region closest to the cell membrane, an interphase, and the region of bulk water. The water closest to the membrane will be closely associated with lipid headgroups, proteins, and other membrane components. The associated water will undergo restricted motion, similar to ice, and its orientation will be determined by its location. Water associated with negatively charged groups will have protons facing towards the group, while the opposite would be true for water associated with positively charged groups. This layer will be a 1−2 water molecules thick, depending on the particular region.

The interphase will be 1−3 nm thick and will be more disorganized and allow for a greater range of motion than the closely associated layer. The molecules will have preferences in orientation based on the local membrane components, but will not have fixed orientations. Exchange between this layer and the bulk layer will occur regularly.

The bulk layer will be typical bulk water, with its regular structures and isotropic tumbling.

A.19.3   While the best model depends on the question being investigated, the Stern model would provide a more accurate model of the interphase because it considers molecular size and non-electrostatic molecular absorption, both of which are very important to biological systems. For example, proteins control the flow of ions through membranes based on size, so if size isn't considered, the interactions leading to flow control would be missing an essential component. Hydrophobic and hydrophilic interactions are also an important part of interphase interactions.

A.19.4   The interaction of the protein with the DNA involves strong ionic interactions. In high ionic strength solutions, the free ions out-compete the DNA for the protein's interactions, releasing the protein from the DNA.

A.19.5   (a) The electric field is calculated by dividing the voltage by the thickness of barrier separating the charges, which in this case is the membrane. This means the change in electric field can be calculated from the change in voltage: $\Delta E = \Delta V / d = (50\,\text{mV} - (-80\,\text{mV}))/(6 \times 10^{-9}\,\text{m}) = 2.167 \times 10^7$ V/m. (b) The capacitance is equal to the total charge divided by the voltage: $C = Q/V$, which can be rearranged to become $C \times V = Q$. Therefore, the change in charges equals the change in voltage times the capacitance: $C \times \Delta V = \Delta Q$. $(1 \times 10^{-6}\,\text{F}) \times (50 - (-80)) \times 10^{-3}\,\text{V} = 1.3 \times 10^{-7}$ C. The charge carried by an electron is 1.602×10^{-19} C, so the number of charges that pass through the membrane to induce the voltage change equals $1.3 \times 10^{-7}\text{C}/1.602 \times 10^{-19}\text{C} = 8.11 \times 10^{11}$.

A.19.6   The electrode surface next to the inner Helmholtz plane is positively charged, suggesting that negatively charged ions will become closely associated with the surface in a way that restricts their motion. This restricted motion will reduce the ion's entropy. Positive ions would be repulsed by the electrode surface, perhaps requiring additional waters around them to help maintain their position. The additional waters would increase the ion's entropy.

A.19.7   The Helmholtz-Perrin model proposes a double layer of ions that exactly cancels the effects of the electrode. A fundamental problem with this proposal is that it does not include the randomizing effects of thermal diffusion as part of its model. See Fig. 20.3.

A.19.8   The Gouy-Chapman model replaces the double layer of the Helmholtz-Perrin model with the diffuse cloud of charge that was more concentrated near the electrode. One of its fundamental problems is that it ignores the effect of the dielectric constant of high-potential fields present at the interface. See Fig. 20.4

A.19.9   The Stern model considers the finite size of ions and the interactions of the ions with the surface, which naturally leads to the Stern layer: a layer of ions closely associated with the electrode surface similar to as found in the Helmholtz-Perrin model. But because the ions have finite size, the charges at the surface cannot completely cancel out the electrode charge; the remaining charge must be canceled by a diffuse layer similar to the Gouy-Chapman model. In dilute solutions, the Stern model behaves similarly to the Gouy-Chapman model; at higher concentrations, it behaves as a mixture of the Helmholtz-Perrin Gouy-Chapman models. See Fig. 20.5.

A.19.10   (a) When $\psi_o = 50$, the superposition will be heavily weighted towards the interactional force, but there will be a definite separation between the two. The crossing point of the superposition will be in the range of $8-10$. (b) When $\psi_o = 10$, the superposition will develop a strong Lennard-Jones character in the range of $r = 5 - 10$, with a definite dip that returns to 0. (c) When $\psi_o = 2.5$, the first portion of the superposition will be almost exactly the same as the Lennard-Jones potential, with the interactional force only slightly tempering it at the further distances. (d) When $\psi_o = 0.5$, the superposition will almost perfectly mimic the Lennard-Jones potential.

A possible mechanism that could decrease the interactional forces between membranes is a change in membrane composition. Altering the lipids and proteins present in one or both membranes could substantially change their electrostatic interactions.

Part IV
Function and Action Biological State Space

Chapter 20
Transport and Kinetics:
Processes Not at Equilibrium

20.1 Questions

Q.20.1 Explain in qualitative but succinct terms the following statement: "Entropy is the primary driving force in transport phenomena." What does this imply?

Q.20.2 What are the four phenomena associated with transport and with which gradient are they each associated? Provide at least one example of each flow from everyday life.

Q.20.3 Name the four types of heat flow (two reversible and two irreversible).

Q.20.4 What is one of the validity requirements for the theory of microscopic reversibility? When did Onsager assume this was the case?

20.2 Answers

A.20.1 There are two reasons why entropy is the driving force. In the first place all transport events happen in a finite period of time. This is necessary for transport to be observed, if it took an infinite amount of time for transport of substances to happen then it would never be discovered and hence could not happen. But this means the transport process can not be reversible and if it is not reversible it must increase the entropy of the universe. Any path that is reversible must have an infinite series of infinitesimal steps. It will take an infinite amount of time to complete. Such a process would never be accomplished in the real world, and hence any real process may only approach reversibility. At best, the entropy of the universe remains constant by having only reversible processes at work; but in reality no reversible process can be completed in a finite time frame, and hence, any real process leads to an increase in entropy.

This chapter from *The Physical Basis of Biochemistry: Solutions Manual to the Second Edition* corresponds to Chapter 21 from *The Physical Basis of Biochemistry, Second Edition*

P.R. Bergethon, K. Hallock, *The Physical Basis of Biochemistry*,
DOI 10.1007/978-1-4419-7364-1_20, © Springer Science+Business Media, LLC 2011

Closely related is the fact that real processes induce scattering as the transport events occur. Scattering is part of every real process and dissipates energy from useful work and into heat. All real processes must scatter aspects of their system (only reversible systems can proceed without disturbing the system state in an infinitesimal series of steps). Scattering is necessary for real transport but scattering means heat production and therefore entropy production.

A.20.2 Diffusion and chemical potential; electrical conduction and electrical potential; heat flow and temperature; fluid flow and pressure. Everyday examples of (1) diffusion: car exhaust expanding past a person on the street, ingested medication (such as ibuprofen and aspirin); (2) Electrical conduction: electric light bulb, power lines, static electricity; (3) Heat flow: steaming cup of coffee, space heater, oven; (4) Fluid flow: tap water, windshield wiper fluid, geyser.

A.20.3 Reversible: Peltier and Thomson; Irreversible: Joule and heat conduction.

A.20.4 It is only valid for deviations from equilibrium where the fluxes are linearly proportional to the forces. He assumed that for processes near equilibrium, equations may be written for the transport process in which the fluxes are linearly proportional to the forces.

Chapter 21
Flow in a Chemical Potential Field: Diffusion

21.1 Questions

Q.21.1 Compared to an aqueous solution at 298 K, would you expect the mean free path of a sodium ion to be longer or shorter in:

(a). Ice at 273 K.
(b). Aqueous solution at 273 K.
(c). Aqueous solution at 373 K.

Q.21.2 Define diffusion.

Q.21.3 Assuming they are in the same solution, do you expect the diffusion coefficient of sucrose or hemoglobin to be higher? Why?

Q.21.4 In a particular three-dimensional solution, the diffusion coefficient of hemoglobin is $7 \times 10^{-11} m^2 s^{-1}$. What's the average distance hemoglobin will diffuse after 1 and 60 s?

Q.21.5 When is the diffusion coefficient's dependence on concentration significant? When is it dependent on the size and shape of the molecule?

Q.21.6 Why does a denatured protein typically have a higher viscosity factor than a folded protein?

21.2 Answers

A.21.1 As an initial assumption, we'll assume that the radius of water and ions do not change significantly over the temperatures being tested. With that in mind, we can use the equations related to the mean free path to see how it relates to temperature.

$$l = \frac{\langle u \rangle}{z}, \langle u \rangle = \left(\frac{8RT}{\pi M} \right)^{1/2}, z = 4\sqrt{\pi}\frac{N}{V}\sigma^2 \left(\frac{RT}{M} \right)^{1/2}$$

$$l = <u>/z \propto 1/(\sigma^2 \times N/V).$$

This chapter from *The Physical Basis of Biochemistry: Solutions Manual to the Second Edition* corresponds to Chapter 22 from *The Physical Basis of Biochemistry, Second Edition*

P.R. Bergethon, K. Hallock, *The Physical Basis of Biochemistry*,
DOI 10.1007/978-1-4419-7364-1_21, © Springer Science+Business Media, LLC 2011

σ is the diameter of the particles, N is the number of particles, and V is the volume. Notably, temperature doesn't play a direct role in determining the mean free path, so assuming that the molecular size doesn't change with temperature, (b) and (c) will have identical mean free paths, although the frequency of collisions will change with temperature. Because (a) involves a phase change, it is more complicated. If the diameter of the particles is identical between the phases, the only thing that changes is density (N/V). At 1 atm, ice is less dense than water and decreased density increases the mean path length. The frequency of collisions would be substantially smaller in ice compared to a liquid, but the mean free path would be longer.

A.21.2   The spontaneous movement of molecules from an area of high concentration to an area of low concentration due to thermal motion.

A.21.3   Sucrose is a smaller, lighter molecule so its diffusion coefficient should be higher.

A.21.4   Use the three-dimensional formula of Equation 22.35, which is described in the text following the equation. sqrt($< x^2 >$) = sqrt($6\,Dt$). $< x >=$ sqrt($6 \times 7 \times 10^{-11} m^2 s^{-1} \times 1s$) = $20.3\,\mu m$, $< x >=$ sqrt($6 \times 7 \times 10^{-11} s^{-1} \times 1s$) = $157.5\,\mu m$.

A.21.5   The diffusion coefficient varies with concentration when the interactions between solutes and solvent are notable. This would be reflected in an activity coefficient that varies in that range of concentrations. The size and the shape of the molecule always impact the diffusion coefficient because those impact the drag a molecule feels moving through a solution.

A.21.6   A folded protein is more spherical than a denatured protein and a sphere is expected to have a lower viscosity factor than non-spherical shapes. This is discussed in more detail in Chapter 27 and shown clearly in Fig. 27.11.

Chapter 22
Flow in an Electrical Field: Conduction

22.1 Questions

Q.22.1 Determine the equivalent conductivity at infinite dilution of the following solutions using Kohlrausch's law of independent migration.

    (a) NaCl
    (b) KCl
    (c) H_2O
    (d) $CaCl_2$
    (e) $Ca(OH)_2$

Q.22.2 When most people lie or make an untrue statement their sympathetic nervous system becomes active (fight or flight response). This leads to increased sweat secretion and thus decreased skin resistance. In a lie detector, a current is passed between two silver chloride electrodes held at 5 v. A current of 4.2 mA is measured when the subject gives his name. Later in the interview he lies and a current of 8.9 mA is measured. What is the skin resistance in the (a) honest and (b) dishonest state?

Q.22.3 The sodium ion is essential to cell signaling in excitable cells. It moves across the charged cell membrane which is an environment of intense electrical fields. Calculate the drift velocity for sodium under the field condition of the transmembrane potentials before depolarization [−80 mV], at full discharge [+50 mV] and during hyperpolarization [−90 mV]. Assume the ion is unhydrated. Use the properties of bulk water at 310 K as necessary.

Q.22.4 What will happen to the drift velocities for the ions in Q.22.3 when the ions are hydrated?

Q.22.5 What are the primary differences between a true and potential electrolyte?

Q.22.6 Does temperature alter the conductivity of ionic solutions? Why or why not?

Q.22.7 Using the data from Fig. 23.6 and Table 23.1, estimate the equilibrium constant of KCl for 0.1, 0.01, and 0.001 M solutions.

This chapter from *The Physical Basis of Biochemistry: Solutions Manual to the Second Edition* corresponds to Chapter 23 from *The Physical Basis of Biochemistry, Second Edition*

P.R. Bergethon, K. Hallock, *The Physical Basis of Biochemistry*,
DOI 10.1007/978-1-4419-7364-1_22, © Springer Science+Business Media, LLC 2011

Q.22.8 What is the electrophoretic effect? Does it increase linearly with concentration? Assuming everything else could be kept constant, how would increasing the ion's charge impact this effect? Is assuming everything is constant a good assumption?

Q.22.9 Why is the conduction of protons in water much higher than expected if it were a regular ion?

22.2 Answers

A.22.1 Use $\Lambda° = v_+\lambda° + v_-\lambda°$. v represents a molar equivalent of the electrolyte. The $\Lambda°$ will be in $\Omega^{-1}m^2equiv^{-1} \times 10^{-4}$ if ionic conductivities of dimension $\lambda° = \Omega^{-1}m^2 \times 10^{-4}$ are used. Use table 23.2 for values and compare results to Table 23.1 or to standard tables.

(a) $[Na^+](50.11) + [Cl^-](76.34) = 126.45$
(b) $[K^+](73.50) + [Cl^{-1}](76.34) = 149.84$
(c) $[H^+](349.80) + [OH^-](197.60) = 547.40$
(d) $[\frac{1}{2}Ca^{2+}](59.50) + [Cl^-](76.34) = 135.84$
(e) $[\frac{1}{2}Ca^{2+}](59.50) + [OH^-](197.60) = 257.1$

A.22.2 Use Ohm's Law to make the calculations: $R = \frac{E}{I}$. E is in volts and I is in amperes. (a) 1200 ohms; (2) 562 ohms

A.22.3 Use $v = \dfrac{z_i e_o E}{6\pi r\eta}$. In order to calculate the electric field E, assume that the membrane across which the transmembrane voltage is measured is 5 nm. We must also assume the viscosity of water, which we assume to be bulk water at 310 K

The first calculation is the electric field:

$$\text{at rest: } \frac{-0.080\,V}{5 \times 10^{-9}m} = -1.6 \times 10^7 V \cdot m^{-1}$$

$$\text{at full depolarization: } \frac{+0.050V}{5 \times 10^{-9}m} = +1.0 \times 10^7 V \times m^{-1}$$

$$\text{at hyperpolarization: } \frac{-0.090\,V}{5 \times 10^{-9}m} = -1.8 \times 10^7 V \times m^{-1}$$

For Na^+ at rest we would write:

$$v = \frac{(1)(1.602 \times 10^{-19}C)(-1.6 \times 10^7 V \times m^{-1})}{6\pi \left(95 \times 10^{-12}m\right)\left(7 \times 10^{-4}kg \times m^{-1} \cdot s^{-1}\right)}$$

$$= 2.046 \times s^{-1} \text{ towards the cell interior}$$

At depolarization:

$$v = \frac{(1)(1.602 \times 10^{-19}C)(+1.0 \times 10^7 V \times m^{-1})}{6\pi \left(95 \times 10^{-12}m\right)\left(7 \times 10^{-4}kg \times m^{-1} \cdot s^{-1}\right)}$$

$= 1.279$ m $\times$ s^{-1} away from the cell interior

At hyperpolarization:

$$v = \frac{(1)(1.602 \times 10^{-19}C)(-1.8 \times 10^7 V \times m^{-1})}{6\pi \left(95 \times 10^{-12}m\right)\left(7 \times 10^{-4}kg \times m^{-1} \cdot s^{-1}\right)}$$

$= 2.302$ m $\times$ s^{-1} towards the cell interior

A.22.4 Solvation will cause the radius to increase proportional to the degree of water association. The hydrodynamic radius is always larger than the crystallographic radius and the drift velocity will be reduced in magnitude. The direction of the drift would be unaffected.

A.22.5 When melted, a true electrolyte is a liquid ionic conductor, and it can create a liquid ionic conductor when dissolved in solvent. A potential electrolyte must interact with a solvent, e.g. water, to create a liquid ionic conductor; it does not form one on its own even when melted.

A.22.6 See Fig. 23.4. At higher temperature, the average diffusion rates are higher, allowing for a more rapid transfer of charge through the solution.

A.22.7 Since we do not have exact values of equivalent conductivity at the desired concentrations, we need to estimate them based on Fig. 23.6. This is best done using a ruler. From Table 23.1, at infinite dilution KCl's equivalent conductivity at infinite dilution is: $\Lambda^\circ = 149.86 \times 10^{-4}\Omega^{-1}m^2equiv^{-1}$; The following were estimated from Fig. 23.6. Your answers may differ a little. [0.1]: $\Lambda \approx 129 \times 10^{-4}\Omega^{-1}m^2equiv^{-1}$, $K = 0.53$; [0.01], $\Lambda \approx 141 \times 10^{-4}\Omega^{-1}m^2equiv^{-1}$, $K = 0.15$; [0.001], $\Lambda \approx 147.5 \times 10^{-4}\Omega^{-1}m^2equiv^{-1}$, $K = 0.069$;

A.22.8 The electrophoretic effect is the increase of viscous force due to the interaction of ionic clouds. It is not linear and it increases as the square root of concentration. Increasing the charge would increase the electrophoretic effect since the strength of ion-ion interactions depends on charge (see Equation (22.23)). No, it's not a good assumption. Changing the charge would change the effective radius of the ion, among other things.

A.22.9 Instead of being physically transported through water, a proton is passed through the hydrogen-bonded lattice from molecule to molecule. The net effect results in the movement of charge, but only hydrogen bonds need to be formed and broken so that transfer rate is much higher than would be expected based on diffusion.

Chapter 23
Forces Across Membranes

23.1 Questions

Q.23.1 What is the Gibbs-Donnan potential? What would happen if the charge on the proteins were switched off? Based on this, suggest a possible mechanism for a cell to modulate the Gibbs-Donnan potential.

Q.23.2 What is a diffusion potential? Under what circumstances does it occur? Propose a membrane that might create a diffusion potential.

Q.23.3 (a) What are the assumptions in the Goldman-Hodgkin-Katz constant field equation? (b) Under what circumstances does the equation tend to fail? (c) Choose one assumption and hypothesize a specific situation that it might not be applicable.

Q.23.4 Consider Fig. 24.4 in the text. Explain the cytoplasmic, membrane, and extracellular components of the above electrostatic profile.

Q.23.5 There are six barrier regions to water traversing a phospholipid bilayer; which region has the highest activation energy and why?

23.2 Answers

A.23.1 The Gibbs-Donnan potential is an electrochemical gradient with offsetting electrical and chemical gradients across the membrane. See Fig. 24.1. There is a persistent electrical gradient driving anions from side B to A, and cations from A to B. The chemical potential gradient drives them in opposite direction. Without the proteins creating a charge imbalance, the ions and water would exchange across the membrane until they reach equilibrium. When $\Delta G = 0$, the system is at equilibrium and the concentrations won't change. A cell could alter the concentrations of proteins and neutralize charges by protein-molecule interactions, pH alterations and

This chapter from *The Physical Basis of Biochemistry: Solutions Manual to the Second Edition* corresponds to Chapter 24 from *The Physical Basis of Biochemistry, Second Edition*

P.R. Bergethon, K. Hallock, *The Physical Basis of Biochemistry*,
DOI 10.1007/978-1-4419-7364-1_23, © Springer Science+Business Media, LLC 2011

by altering other impermeable charged species like lipids to modulate the Gibbs-Donnan potential.

A.23.2 A diffusion potential occurs when ion solutions of different concentrations are connected by a semi-permeable membrane that allows free exchange of water and ions. The positive and negative ions must also have different mobilities moving through the membrane. Their difference in mobility leads to a separation of charge during diffusion, creating a diffusion potential. A membrane with scattered positive or negative charges could create a diffusion potential because it would attract the opposite ions, slowing them down as they passed through it.

A.23.3 (a) (1) the partial permeability of the membrane to charged species, (2) the uniformity of charge distribution across the membrane, (3) the net equality of charge flux across the membrane, (4) the applicability of the Donnan equilibrium for each ionic species to which the membrane is permeable enough to permit free distribution across the membrane, and, (5) the absence of an electrogenic pump and its resultant driving force.

(b) The Goldman-Hodgkin-Katz constant field equation holds for most mammalian cells around physiological conditions, but tends to fail when such cells are suspended in non-physiologic buffers.

A.23.4 The cytoplasmic electrostatic profile is created by surface charges, with more negatively charged proteins and fewer positive ions then the extracellular side. The membrane portion the electrostatic profile is strictly due to the transmembrane potential; the membrane may be treated as a dielectric continuum with a dielectric constant of 8. The electrostatic profile due to the extracellular surface of the plasma membrane assuming fewer negatively charged proteins and more positive ions than the cytoplasmic side.

A.23.5 The non-aqueous hydrophobic middle of the lipid bilayer has the highest activation energy because the water has stronger interactions in bulk water where it can form hydrogen bonds, or at the bilayer surface where it can interact with charges found on the phospholipid headgroups. The loss of these favorable interactions creates a large activation energy for passing through the hydrophobic middle of a phospholipid bilayer.

Chapter 24
Kinetics – Chemical Kinetics

24.1 Questions

Q.24.1 From Table 25.2 determine the $\frac{k_{cat}}{K_M}$ for the following enzymes. What is the likely rate limiting step in the catalytic process.
Chymotrypsin; Carbonic anhydrase; Catalase; Penicillinase; Lysozyme

Q.24.2 What is the effect on the rate of a reaction at 273 K with an activation energy of 25 kJ mol^{-1} when the temperature increases 1°? 5°? 10°? 20°? 50? 100°?

Q.24.3 Calculate the change in rate constants for the following activation energies with the accompanying temperature changes:

(a) 50 kJ mol^{-1} from 0° C to 10°C
(b) 5 kJ mol^{-1} from 0° C to 10°C
(c) 500 kJ mol^{-1} from 0° C to 10°C

Q.24.4 A reaction has the following rate constants:

$$A < - - - > B \quad K_1 = 10^{-5}s^{-1}$$

$$K_{-1} = 10^{-8}s^{-1}$$

(a) What is the equilibrium constant for the reaction?
(b) Which way will the reaction go if $[A] = 2.6 \times 10^{-6}$ M and $[B] = 1.5 \times 10^{-4}$ M?
(c) The reaction will proceed in which direction if $[A] = 1.5 \times 10^{-1}$ M and $[B] = 1.5 \times 10^{-1}$ M?

Q.24.5 The half life of ^{14}C is 5568 years. This isotope is produced in the atmosphere by the interaction between CO_2 and cosmic rays. There is a steady state concentration of ^{14}C in the air. Plants that use the $^{14}CO_2$ incorporate

---

This chapter from *The Physical Basis of Biochemistry: Solutions Manual to the Second Edition* corresponds to Chapter 25 from *The Physical Basis of Biochemistry, Second Edition*

P.R. Bergethon, K. Hallock, *The Physical Basis of Biochemistry*,
DOI 10.1007/978-1-4419-7364-1_24, © Springer Science+Business Media, LLC 2011

the isotope into their molecular structure with no further additions or sub-tractions except by radioactive disintegration. If a sample of wood from a tree is found to have the ^{14}C content of 86% compared to atmospheric ^{14}C (assume that this number is constant), what is the age of the tree?

Q.24.6   The following data is collected for an enzyme:

Substrate concentration (mmols)	Initial velocity (μmols s^{-1})
1.716	0.416
3.432	0.728
5.20	1.040
8.632	1.560
13.00	2.028
17.316	2.392
25.632	3.01

(a)   What is the K_m for this enzyme?

(b)   What is V_{max}?

Q.24.7   The binding of a substrate to an enzyme increases its reaction velocity by a factor of 1000.

(a)   What is the change in the free energy of activation?

(b)   Another enzyme accelerates the same reaction by a factor of 10,000; what is the energy of activation of this enzyme-substrate complex?

Q.24.8   What are some of the questions asked by chemical kinetics?

Q.24.9   Under certain experimental conditions, the conversion of ethanol into acetaldehyde by alcohol dehydrogenase has a rate of 25 mmol L^{-1} s^{-1}. If we begin with 1 M ethanol and no acetaldehyde at $t = 0$, how much acetaldehyde is created after 13 s? Assume it's a zero-order reaction. Is this a good assumption?

Q.24.10   Under certain experimental conditions, the half-life of H_2O_2 is 28 h. Find its rate of decay assuming it's a first-order reaction. At 40 h, find out how much H_2O_2 remains in a solution that began with 4 mmolar H_2O_2 after estimating the amount based on the half-life.

Q.24.11   The conversion of a small strip single-stranded DNA to double-stranded DNA follows second-order kinetics with a rate constant of 0.17 mmolar^{-1} min^{-1}. If the initial concentration of single-stranded DNA was 1.2 mmo-lar, how much DNA remains after 5 min?

Q.24.12   What does the elucidation of an elementary reaction include?

Q.24.13   What is an activated complex? Where on a potential energy surface does it occur?

Q.24.14   What is the transmission coefficient? Discussing this in a macroscopic manner, assume that a truck is parked at the top of a steep hill and after numerous experiments, its transmission coefficient was determined to be 0.25; what does that suggest about the truck's mass distribution?

Q.24.15 If V_{max} for catalase $= 1.87 \times 10^5$; what is its concentration of active sites?

Q.24.16 What three locations can be used to control enzymatic activity?

Q.24.17 What are the differences between competitive, noncompetitive, and uncompetitive inhibition?

24.2 Thought Review

There are four dominant methods of chemical catalysis that occur in biological systems.

(1) Entropic or proximity catalysis
(2) Electrophilic and nucleophilic attack
(3) General acid-base catalysis
(4) Strain and distortion

You should be able to select an example of each and discuss in terms of the potential energy surface the forces involved in the catalytic mechanism.

24.3 Answers

A.24.1 k_{cat}/K_M: Chymotrypsin $= 2 \times 10^4$; Carbonic anhydrase $= 7.5 \times 10^7$; Catalase $= 3.89 \times 10^7$; Penicillinase $= 4 \times 10^7$; Lysozyme $= 8.33 \times 10^4$. Diffusion is the most likely limiting factor for Carbonic anhydrase, Catalase, and Penicillinase because their k_{cat}/K_M is on the order of 10^7. Chymotrypsin and lysozyme are most likely limited by something other than diffusion.

A.24.2 Use $\ln k' - \ln k = -\frac{E_a}{RT'} + \frac{E_a}{RT}$. Since A will be the same for both reaction it can be ignored in this solution. This equation can be rearranged to $\ln \frac{k'}{k} = \frac{E_a}{R}(\frac{1}{T} - \frac{1}{T'})$. The rate increases for each of the increasing temperatures respectively: 0.4%, 1.9%, 3.9%, 7.7%, 18%, 34.3%.

A.24.3 The Arrhenius equation can be used to solve the change in this case.

$$\ln k_1 = \ln A - E_a/RT_1; \quad \ln k_2 = \ln A - E_a/RT_2.$$

$$\ln k_2 - \ln k_1 = (\ln A - E_a/RT_2) - (\ln A - E_a/RT_1) = E_a/RT_1 = E_a/RT_2$$

$$\ln(k_2/k_2) = (E_a/R)(1/T_1 - 1/T_2)$$

$$R = 8.314\,\text{J K}^{-1}\text{mol}^{-1}; \quad T_1 = 273\,\text{K}; \quad T_2 = 283\,\text{K};$$

Solving the above equation, we get the following results:

(a) $E_a = 50,000\,\text{J mol}^{-1}$, $\ln(k_2/k_1) = 0.778$, $k_2/k_1 = 2.18$.
(b) $E_a = 5000\,\text{J mol}^{-1}$, $\ln(k_2/k_1) = 0.0778$, $k_2/k_1 = 1.08$.
(c) $E_a = 500,000\,\text{J mol}^{-1}$, $\ln(k_2/k_1) = 7.78$, $k_2/k_1 = 2400$.

The relative rate of the highest activation energy reaction (c) increases substantially more than the lowest activation energy (b) because the rate is negligible in (c), so any increase in temperature results in a relatively large change, while low activation energy reactions have appreciable rates that are only increased a little with a small increase in temperature.

A.24.4 (a) 1000; (b) A $\longrightarrow$ B; (c) B $\longrightarrow$ A

A.24.5

$$\frac{[^{14}C_{\text{sample}}]}{[^{14}C_{\text{atmosphere}}]} = 0.86$$

$$\ln\frac{[^{14}C_{\text{sample}}]}{[^{14}C_{\text{atmosphere}}]} = \frac{0.693}{t_{1/2}}t$$

$$\ln[0.86] = -\frac{0.693}{5569}$$

$t = 1212$ years old

A.24.6 Graph the data to find the line. Solution to the line is $y = .0038x + .000216$. $V_{\text{max}} = 4.6\,\text{mmol s}^{-1}$ and $K_m = 1.72\,\text{mmol}$.

A.24.7 (a) 10^3 increase in v is associated with a ΔG of $-17.11\,\text{kJ mol}^{-1}$.
       (b) 10^4 increase in v is associated with a ΔG of $-22.84\,\text{kJ mol}^{-1}$.

These solutions are found by:

$v \propto [S^*] \propto \Delta G^*$ because the transition state is in equilibrium with S.

$$[S^*] = [S]e^{-\Delta G^*/Rt} \text{ and } v = v[S^*]e^{-\Delta G^*/RT}$$

A.24.8 How fast is the reaction getting to where it is going? What is the reaction path for the reaction? What are the rates of each of these steps?

A.24.9 Zero-order reactions are discussed in section 25.5.1. $[A_0] = 0$ for acetaldehyde. The rate for the conversion of ethanol into acetaldehyde, i.e. the loss of ethanol, is 25 mmol L^{-1} s^{-1}, so the rate of the creation of acetaldehyde is the opposite: -25 mmol L^{-1} s^{-1}. Rearranging Equation 25.9, we get $[A] = [A_0] - kt = 0 - (-25\,\text{mmol L}^{-1}\text{s}^{-1}) \times 13\,\text{s} = 325\,\text{mmol L}^{-1}$. The reaction could behave as a zero-order reaction if the concentration of ethanol is much higher than alcohol dehydrogenase, in which case, the rate of conversion would be independent of ethanol concentration. This situation is often referred to as a pseudo zero-order reaction.

A.24.10 The first step is to determine the rate from the half-life. From the text below equation 25.14, half - life $= 0.693/k$. $k = 0.693/28\,\text{h} = 0.02475\,\text{h}^{-1}$. 40 h is in-between one half-life (28 h) and two half-lives (56 h), so you

expect the final concentration to be between 0.5 and 0.25 the original concentration. In this case, the concentration should be between 1 mmolar and
2 mmolar. With the rate calculated, we can use equation 25.13 for the rest.
$[A] = [A_0] \times e^{-kt} = [4\,\text{mmolar}] \times e^{-(0.2475\,\text{h}- \times 40\,\text{h})} = 1.49\,\text{mmolar}$.

A.24.11 Rearranging equation 25.18, $1/[A0] + kt = 1/[A]$. $k = 0.17\,\text{mmolar}^{-1}$
min^{-1}; $t = 5\,\text{min}$; $[A0] = 1.2\,\text{mmolar}$. $1/[A] = 1/(1.2\,\text{mmolar}) + 0.17\,\text{mmolar}^{-1}\,\text{min}^{-1}\,\text{min} = 1.683\,\text{mmolar}^{-1}$. $[A] = 0.59\,\text{mmolar}$.

A.24.12 (1) A description of the type and number of molecules that participate in
the reaction. (2) The energies of and the accessibility to the activated or
transition states formed in the path of the elementary reaction.

A.24.13 An activated complex is when two or more molecules need to be placed in
an activated state for a reaction to proceed. The activated complex occurs
at a local maximum on the potential energy surface.

A.24.14 It is the probability that the transition state will continue on to become
products; it is usually a value between 0.5 and 1. A transmission coefficient
of 0.25 means the truck is more likely to roll backwards than forwards,
suggesting there is substantially more mass in the back of the truck than
the front of the truck.

A.24.15 If V_{max} for catalase $= 1.87 \times 10^5$; what is its concentration of active sites?
Answer: Using Equation 25.66 and Table 25.2, $[ET] = (1.86 \times 20^5)/(4.67 \times 10^4) = 3.98 \approx 4$.

A.24.16 (1) The active site, (2) a regulatory site that alters the control surface of
the active site, and (3) the access regions to the active site that allow the
substrates to arrive or the products to leave the active site region.

A.24.17 Competitive inhibition is a molecule that mimics the transition by occupying the catalytic site, but the molecule is unreactive; this increases
K_M. Noncompetive inhibitors bind to a site on the enzyme other than
the catalytic site, decreasing V_{max} without affecting K_M. Uncompetitive
inhibition affect both K_M and V_{max}.

Chapter 25
Bioelectrochemistry – Charge Transfer in Biological Systems

25.1 Questions

Q.25.1 The blood vessel can be modeled as a tube with an electrolyte traveling through it under a hydrostatic pressure. The ζ potential of the vessel is -200 mV; the pressure is 16×10^3 Pa; blood viscosity is 3.5×10^{-3}kg - m^{-1}s^{-1}; the conductivity is 0.67 S - m^{-1}; use a dielectric constant of 10, the permittivity of free space is 8.85×10^{-12}F - m^{-1}.

   (a) What is the streaming potential under these conditions?
   (b) What happens to the streaming potential if the patient is hypertensive and the mean arterial blood pressure doubles?
   (c) What assumptions are made in this analysis?
   (d) Are these assumptions valid?

Q.25.2 The zeta potential depends on the surface charge of the vessel. Endothelial cells have a net negative surface charge. However in cases of vascular injury the endothelial cells are stripped off to reveal the sub-endothelial lining which contains a high content of collagen. Collagen contains a large percentage of lysine making this new surface much more basic in nature. Given the blood and tissue pH of 7.4, what happens to the streaming potential with endothelial injury?

Q.25.3 Explain how electroneutrality is maintained in an electrical circuit that has an electrolytic conductor as a circuit element. Be sure to address the three general aspects of the circuit's behavior, that is, electronic, ionic, and electrodic.

Q.25.4 Considering only $\Delta G°$ and λ, when is the Frank-Condon factor maximized? What is λ and what does it represent?

Q.25.5 Considering the flow of electrons in the photosynthetic reaction center of *R. viridis*, why does the reaction proceed forward despite the back reaction being favored thermodynamically?

This chapter from *The Physical Basis of Biochemistry: Solutions Manual to the Second Edition* corresponds to Chapter 26 from *The Physical Basis of Biochemistry, Second Edition*

Note: An important aspect of bioelectrochemial analysis are the processes of electrophoresis and electrical transport (conduction). Questions specific to those topics
but appropriate to this chapter are found in Chapters 23 and 27.

25.2 Answers

A.25.1 (a) In order to calculate the steaming potential use $E_s = \frac{\zeta \varepsilon_0 \varepsilon P}{\eta \kappa}$.
Substituting the values given in the problem gives the streaming
potential in volts: $120 \, \mu V$.

   (b) The streaming potential is proportional to the pressure and doubling
the pressure will double the streaming potential.

   (c) There are a number of assumptions made in this mathematical model
of a complex biological system. Certain basic assumption include laminar flow of the fluid in the capillary, reasonable Debye length of the
electrical double layer with respect to the surface topology of the cap-
illaries, no surface conduction along or through the capillary. The flow
of liquid is assumed to be constant and non-turbulent.

   (d) These may not be fulfilled in this simple model since blood is a
non-newtonian fluid that is not always laminar in flow and surface
conduction is likely. In addition, blood flow is pulsatile and this formula is not corrected for pulsatile flow. Turbulence often appears in
blood flow, especially at bifurcations of vessels and this may also need
correction.

A.25.2 At the blood pH, the lysine amino acid groups are ionized and the
exposed surface will be more positive. This will change the zeta potential at the blood-vessel interface and may even switch the sign of the zeta
potential.

A.25.3 Electroneutrality must be maintained in all aspects of the circuit. In the
electronic portions of the circuit the charge carriers are largely electrons
though in semiconductors, positive charge carriers called holes can carry
current. Within the electrolyte itself the charge will be carried by ions moving through the solvent down their electrical field gradient. The exchange
from the electronic charge carriers to ionic charge carriers will occur at
the electrode interface and here electrochemical reactions will be required
to transfer charge through coupled oxidation-reduction reactions. The net
charge at the electrodes must be balanced and electroneutral just as in all
other portions of the overall circuit or current flow will be opposed and
current flow will cease.

A.25.4 The Frank-Condon factor has a maximum when $\Delta G° = \lambda$. λ is the
reorganization energy and it represents the energy needed to distort the
product state into the same shape as the initial state without electron transfer
occurring.

A.25.5 This charge-separation reaction demonstrates an important difference between the kinetics of a reaction and the thermodynamics of a reaction. Thermodynamics shows us where a reaction would like to go given an infinite of time, but most reactions happen much faster than that. In this case, the large $-\Delta G°$ reduces that rate of the back reaction so much that the forward reaction is kinetically favored, so the reaction proceeds forward.

Part V
Methods for the Measuring Structure and Function

Chapter 26
Separation and Characterization of Biomolecules Based on Macroscopic Properties

26.1 Questions

Q.26.1 Calculate the work performed by a battery delivering 100 milliamps at 9 volts for 2 hours. Express the answer in (a) joules, (b) calories and (c) watt-hours.

Q.26.2 What is the work performed by a 50 kg gymnast who performs a lift on the rings and is elevated 1.2 m in the air?

(a) How many calories must be supplied to the muscles?
(b) Assuming 100% efficiency in energy extraction from sugar and given that there are 4 kcal/gram of sugar (5 g in a teaspoon, how many teaspoons of sugar should be consumed at lunch to make the lift?

Q.26.3 The difference between normal hemoglobin A and the sickle-cell hemoglobin mutant protein is a single amino acid replacement of glutamate with valine in the β chains. (Each hemoglobin molecule is composed of 2α and 2β chains). The mobility of these two proteins can be measured as $+0.3 \times 10^{-9} m^2/s$ - V and $-0.2 \times 10^{-9} m^2/s$ - V.

(a) Match the mutant versus native protein with its respective mobility.
(b) On an isoelectric focusing gel which protein will be found at the more acidic position in the pH gradient?

Q.26.4 On average, the addition of $SDS[H_3C - (CH_2)_{10} - CH_2O - SO_3^- Na^+]$ to a denatured protein leads to a stoichiometry of 2 amino acids/1 SDS molecule. Estimate the SDS associated charge on the following proteins following denaturation. [A reasonable rule of thumb is to assign each amino acid a weight of 100 daltons.]

(a) Albumin (67.5 kD)
(b) Ribonuclease A (12.4 kD)

This chapter from *The Physical Basis of Biochemistry: Solutions Manual to the Second Edition* corresponds to Chapter 27 from *The Physical Basis of Biochemistry, Second Edition*

P.R. Bergethon, K. Hallock, *The Physical Basis of Biochemistry*, DOI 10.1007/978-1-4419-7364-1_26, © Springer Science+Business Media, LLC 2011

   (c) Fibrinogen (330 kD)

   (d) Hemoglobin (68 kD)

   (e) Myoglobin (18 kD)

Q.26.5 You have isolated a new protein (X) from cellular fragments. You run an isoelectric focusing gel to help characterize the protein. Three standards (A, B, C) are run along with the new protein. The standard's pIs are given below.

Standards name	pI
b-lactoglobulin	5.2
Cytochrome c	10.6
Albumin	4.8

After running your gel you realize that you have lost the key to your standards and so run a separate gel to figure out their order. The two gels (one standards only and one standards + unknown) have the following order:

Standards:

```
              A   B                      C
pH gradient:    3------------------------------------------------>12
```

Standards + unknown

```
              A   B              X  C
pH gradient:    3------------------------------------------------>12
```

   (a) Make a new key for the standards.

   (b) What conclusions about your unknown protein can you make from this experiment?

   (c) What further experiments could you now perform to further characterize the physical nature of the protein?

Q.26.6 Proteins with subunit chains linked by cysteine-cysteine disulfide bonds (RS-SR) can be reduced to sulfhydryl (R-SH HS-R) containing unlinked chains by the addition of β-mercaptoethanol or dithiothreitol. These reduced proteins will then run independently according to their molecular weights on an SDS denaturing gel. The following pattern is obtained for a mixture of proteins:

Non-reducing lane   **Reducing lane**

a ------------   ------------

b ------------

c ------------   ------------

                                                   ----

d ------------   ------------

Running buffer   ------------   ------------

(a) Which of the proteins in the original sample are linked by disulfide bonds?
(b) If the molecular weight of d is 18,000 daltons and a is 130,000 what is the approximate weight of the unknown protein and its fragments.
(c) How many subchains make up the whole protein? How could you support your argument?

Q.26.7 (a) What is Archimedes' principle? (b) Using a cup filled with water, a pan to catch the water, and several small objects, test Archimedes' principle.

Q.26.8 A parachutist jumps out of an airplane and falls towards the ground until he reaches terminal velocity, which in this case is 40.18 m/s. If his mass is 82 kg, what was his drag coefficient (b)? If the parachutist's drag coefficient is suddenly increased by a factor of 10, what might he have done?

Q.26.9 How is sedimentation analysis used to separate macromolecules? What are some of the properties of macromolecules that might determine their frictional coefficients?

Q.26.10 What is the typical name, order of magnitude, and units of the sedimentation coefficient?

Q.26.11 What orientation of a dinner plate would be easier to push through a pan of water? Why?

Q.26.12 Identical amounts of thyroxine transport protein (TTR) are placed in water and gasoline. Identical sedimentation experiments are run on both samples. It's discovered that TTR sediments faster in water; what is a likely explanation of this observation?

Q.26.13 What factors impact protein separation by electrophoresis? What does a scientist typically do if they want to separate proteins by molecular weight only?

Q.26.14 (a) Define partition coefficient ratio, theoretical plate, and retention time. b) Choose a protein property (size, or charge, etc.) and describe the liquid-solid chromatographic technique you would use to separate it. Draw the device, labeling the essential parts needed for separation.

Q.26.15 (a) What does the mass spectrometer use to separate analytes? (b) A researcher hypothesizes that two compounds have different molar masses. She injects them into a mass spectrometer and they both have identical peaks; propose several reasons to explain this result.

26.2 Answers

A.26.1 You may need to review Chapter 6. Determining the electrical work is usually done by calculating the power per second and multiplying by the time the electricity was used. $w_{electrical} = \text{Power} \times \text{time} = -E \times I \times t$

$E = 9v$; $I = 0.100\,\text{amps}$; $= 2\,\text{h}$. $1\,\text{volt} = \text{J/coulomb}$; $1\,\text{ampere} = \text{coulomb/s}$. $4.186\,\text{J} = 1\,\text{cal}$, $1\,\text{J} = 0.2389\,\text{cal}$; $1\,\text{J} = 2.778 \times 10^{-4}\,\text{w - h}$
(a) $w_{electrical} = 9\,\text{V} \times 0.100\,\text{A} \times 2\,\text{h} \times 60\,\text{min/h} \times 60\,\text{s/min} = 9\,\text{J/C} \times 0.100\,\text{C/s} \times 2\,\text{h} \times 60\,\text{min/hr} \times 60\,\text{s/min} = 6480\,\text{J}$. (b) $6480\,\text{J} \times 0.2389\,\text{cal/J} = 1548.1\,\text{cal}$. (c) $6480\,\text{J} \times 2.778 \times 10^{-4}\,\text{w - h/J} = 1.80\,\text{w - h}$

A.26.2 Lifting work is determined by multiplying the mass by the acceleration, which in this case is caused by Earth's gravity, and by the change in height. $w = m \times g \times \Delta h$. $m = 50\,kg$, $g = 9.8\,m/s^2$, $\Delta h = 1.2\,m$. $1\,J = 1\,kg \cdot m^2/s^2$, $1\,J = 0.2389$ calorie. (a) $w = 50\,kg \times 9.8\,m/s^2 \times 1.2\,m = 588\,J$, (b) $588\,J \times 0.2389\,cal = 140.5\,cal$, (c) $140.5\,cal \times 1\,gram/4000\,cal \times 1\,tsp/5\,gram = 7 \times 10^{-3}\,tsp$.

A.26.3 (a) Glutamate is usually negatively charged at physiological pH and valine is neutral. In electrophoresis, the more negatively charged proteins have a stronger attraction for the positive electrode. Hemoglobin A is expected to have the higher mobility ($+0.3 \times 10^{-9}\,m^2/s$ - V). b) At low pH, glutamate is protonated to become glutamic acid, so hemoglobin A will be more acidic.

A.26.4 The problem states that charge is related to the number of amino acids and those can be estimated assuming $0.1\,kD = 1$ amino acid, and we can assume 2 amino acids equal 1 SDS molecule = one negative charge. The total charge added to each protein is then it's molar mass divided by $0.2\,kD$. (a) 337.5, (b) 62, (c) 1650, (d) 340, (e) 90.

A.26.5 (a) A has the lowest isoelectric point, so it is albumin; B has the next lowest so it's β-lactoglobulin, and cytochrome c is C. (b) That it has a pI of about 10. (c) Electrophoresis would provide an estimate of molecular weight. Circular dichroism could provide an estimate of α-helix content. NMR and x-ray crystallography might be able to provide some structural information.

A.26.6 (a) The only protein that is in the non-reducing lane that is absent from the reducing lane is b, so b is the protein linked by disulfide bonds. (b) The non-reduced b is has a lower mass than a, but not by much, while its subunits are equidistant from d. Based on its distance from a, whole b is probably near 100,000 daltons, and given their symmetry around d, its subunits are probably around 30,000 daltons and 10,000 daltons. (c) Two large subunits and four small units would give the approximate weight of the whole protein, but other combinations, such as three large subunits and one small subunit, are possible as well. Investigating the possibilities would require semi-quantitative techniques, perhaps by evaluating the density of the bands in the electrophoretic gel.

A.26.7 (a) The buoyant force acting on an immersed object is equal to the weight of the fluid displaced by the object. (b) Don't make too big of mess!

A.26.8 Rearranging equation 26.6: $b = m \times g/v_t = (82\,kg \times 9.8\,m/s^2)/40.18\,m/s = 20\,kg\,s^{-1}$. Opening the parachute is one way to increase his drag coefficient, hopefully he did just that.

A.26.9 Sedimentation analysis separates macromolecules by applying a force to them while they're in solution. Their different frictional coefficients lead to different rates of sedimentation. Mass, size, and shape. A long floppy protein will sediment slower than a similarly massed compact protein.

A.26.10 Svedberg (S); 10^{-13} s.

A.26.11   Pushing a plate through the water with the thinner side facing the direction of motion would reduce the drag proportionality constant, making it easier to push the plate through the water.

A.26.12   To answer this question, we need to consider what we know about the system. We know that gasoline is less dense than water, which suggests that if everything else is equal, TTR should sediment faster in gasoline. Since it doesn't, something else must change. The problem stipulates that identical experiments were run on the systems, so spinning speed, etc. were all the same so something with the protein must have changed. Most likely, TTR unfolds in gasoline increasing its frictional coefficient, although that is not the only possibility. Gasoline might degrade protein into small pieces, and those smaller pieces would also sediment slower.

A.26.13   The mobility of a macromolecule will depend on the net charge, size, shape and adsorption characteristics. If an estimate of molecular weight is to be performed the common practice is to (1) neutralize the shape differential by denaturing the macromolecule (usually by boiling and reducing all intramolecular disulfide bonds. (2) Neutralize the intrinsic charge differential by covering the macromolecule with a charged amphiphile that gives it a proportional charge per unit length. This is accomplished by mixing the denatured protein with SDS, sodium dodecyl sulfate, a detergent.

(3) Run the sample in an appropriately cross-linked gel to enhance the molecular sieving along with reference molecular weight samples.

A.26.14   (a) The partition coefficient ratio is the ratio of the time the analyte spends in the mobile phase compared stationary phase. Theoretical plates refers to the number of times an analyte partitions between the phases. Retention time is the time required for the analyte to travel past the stationary phase.
(b) Answers will vary.

A.26.15   (a) Mass spectrometry uses the mass-to-charge ratio (m/z) of the gaseous ionized molecular ions. Its separation results in a plot that shows the abundance of a particular molecular fragment as a function of its m/z ratio.
(b) The compounds could have identical masses and charges. One compound could be a multiple of the mass of the other compound, e.g. 2:1, and have the same ratio in charge. The researcher could have mistakenly used the same compound for both tests. Because of the last possibility, the researcher should repeat her experiment to make certain the result is reproducible.

Chapter 27
Determining Structure by Molecular Interactions with Photons: Electronic Spectroscopy

27.1 Questions

Q.27.1 You are studying an unknown compound and find that it is soluble in water and in methanol. You perform UV-visible absorption spectroscopy. In water there is a small adsorption peak at 275 nm that shifts to 290 nm when the same scan is performed in methanol. What is the likely structure and the electronic transition that gives rise to this observation?

Q.27.2 Calculate the concentration of NADH ($e = 6200$ liter - mole^{-1}cm^{-1}) The absorbance is measured at 340 nm in a 1 cm cell to be (a). 0.345; (b) 0.556; (c) 0.145.

Q.27.3 You are studying an ion channel for which you want to know the opening pore dimension. Fortunately there are two and only two sites at the mouth of the pore to each of which a fluorescent label may be attached. You decide to study the dimension with Forster analysis, the efficiency of Forster transfer is given by:

$$\text{Efficiency} = \frac{1}{1 + (r/R_o)^6}$$

You choose a pair of fluorochromes whose characteristic R_o value in water is 27 Å. You measure an energy transfer efficiency of 92%. What is the approximate dimension in the pore.

Q.27.4 A solvent with an ionic strength of 0.15 alters the R_o of fluorochrome pair above to 15 Å because of the effect of the local electric field from the ions on the resonance transfer. You reconstitute your pore assembly and notice that when the conductivity of the membrane increases the Forster efficiency drops to 25%. Explain this effect and describe the state of the environment in the pore in the conducting and non-conducting state.

This chapter from *The Physical Basis of Biochemistry: Solutions Manual to the Second Edition* corresponds to Chapter 28 from *The Physical Basis of Biochemistry, Second Edition*

P.R. Bergethon, K. Hallock, *The Physical Basis of Biochemistry*,
DOI 10.1007/978-1-4419-7364-1_27, © Springer Science+Business Media, LLC 2011

Q.27.5 The ratio of NAD$^+$ and NADH can be used to spectrophotometrically measure the redox potential of a system that includes this pair. All measurements were conducted using an instrument with a 1 cm path length. What is the redox potential of the following samples:

(a) absorption I @260 nm = 0.345; absorption @340 nm = 0.345.
(b) absorption @260 nm = 0.105; absorption @340 nm = 0.345.
(c) absorption @260 nm = 0.965; absorption @340 nm = 0.250.

Q.27.6 You are studying the effect of superoxide anion production by macrophages (one of the white cells in the immune system) on membrane structure. A nitroxide EPR spin label is added to the membrane. EPR monitoring of the spin-label signal is continuous. As the experiment starts the first change noted is the development of hyperfine splitting of the spin-label signal. Account for this change.

Q.27.7 The experiment from above continues. During the continuous monitoring of the EPR signal:

(a) The signal is noted to broaden. How can you interpret this change.
(b) What other methods could be used to prove your explanation for a.

Q.27.8 (a) What is a chemical shift and how is it created?
        (b) Would the carbons in ethylene or dimethyl ether have greater chemical shifts?

Q.27.9 (a) What is J-coupling?
        (b) Predict the J-coupling splitting pattern for chloroethane.

Q.27.10 What is T_1 and what is one way it can be measured?

Q.27.11 What are the three types of magnetism and how do they differ?

Q.27.12 Given the following rates of fluorescence and absorption of several different proteins, calculate the following quantum yields. (a) Fluorescence Rate: 50 s^{-1}, Absorption Rate: 60 s^{-1}; (b) Fluorescence Rate: 33 s^{-1}, Absorption Rate: 42 s^{-1}; (c) Fluorescence Rate: 112 s^{-1}, Absorption Rate: 88 s^{-1}; (d) Fluorescence Rate: 8 s^{-1}, Absorption Rate: 91 s^{-1};. Do the answers all make physical sense? Why or why not?

27.2 Answers

A.27.1 The unknown compound that absorbs in the 275−290 nm range undergoes a bathochromic shift when dissolved a solvent less polar than water. Looking at Table 28.1, there are two possibilities based on the wavelengths: C=O and -S-S-. (Nucleic acids and amino acids have some possible absorbances because they have these functional groups.) C=O is expected to undergo a bathochromic shift when placed in a less polar solvent (Table 28.2), while nothing is included on that same table about the -S-S- transitions. Based on the limited information, C=O is the most consistent with our data, but we can't rule out -S-S-

A.27.2 Calculate the concentration of NADH ($\varepsilon = 6200$ l - mole^{-1} cm^{-1}). The absorbance is measured at 340 nm in a 1 cm cell to be (a) 0.345; (b) 0.556; (c) 0.145.

The Beer-Lambert law allows the calculation of concentrations based on relative absorbances. It's only accurate at dilute concentrations, so we'll assume the above measurements were taken at dilute concentrations and then revisit that assumption after doing the calculations.

$A = \varepsilon lc$; $l = 1$ cm, $\varepsilon = 6200$ l - mole^{-1} cm^{-1}, $A =$ reported absorbances.

Taking the logarithm of both sides and rearranging, we get: $c = A/\varepsilon l$

(a) $c = A/\varepsilon l = 0.345 \times (1/6200$ l - mole^{-1}cm$^{-1} \times 1$ cm$) = 55.6$ μM

(b) $c = A/\varepsilon l = 0.556 \times (1/(6200$ liter - mole^{-1}cm$^{-1} \times 1$ cm$) = 89.7$ μM

(c) $c = A/\varepsilon l = 0.145 \times (1/6201$ - mole^{-1}cm$^{-1} \times 1$ cm$) = 23.4$μ M

The concentrations are relatively low, so the dilute solution assumption was reasonable.

A.27.3 Rearranging the Forster transfer equation, we have:

(Efficiency) $(1 + (r/Ro)^6) = 1$

Efficiency $\times (r/Ro)^6 = 1 -$ Efficiency

$r^6 = Ro^6 \times ((1 -$ Efficiency)/Efficiency)

$r = (Ro^6 \times ((1 -$ Efficiency)/Efficiency))$^{1/6} = (27$ Å$)^6 \times ((1 - 0.92)/0.92))^{1/6}$

$r = 18$ Å

A.27.4 In the non-conducting state the pore likely contains free water. In the conducting state ions are in the channel.

A.27.5 The ratio of NAD$^+$ and NADH can be used to spectrophotometrically measure the redox potential of a system that includes this pair. What is the redox potential of the following samples:

(a) absorption @260 = 0.345; absorption @340 = 0.345.

(b) absorption @260 = 0.105; absorption @340 = 0.345.

(c) absorption @260 = 0.965; absorption @340 = 0.250.

As shown on Table 28.1, the 260 nm absorbance is from the adenine in nicotinamide adenine dinucleotide (NAD$^+$ and NADH). Therefore, the absorbance at 260 nm measures the combined concentrations of NAD$^+$ and NADH. According to Table 28.1, the reduced molecule (NADH) absorbs at 340 nm, while NAD$^+$ has no absorbance, so the concentration determined from that wavelength will equal the concentration of NADH. The extinction coefficients at 260 nm is 13,400 l-mole^{-1} cm^{-1} and at 340 nm is 6200 l-mole^{-1} cm^{-1} (Table 28.1).

NADH Concentration: (a) 25.6 μmole, (b) 1.8 μmole, (c) 23.4 μmole,
NAD$^+$ Concentration: (a) 0.1 μmole, (b) 20.6 μmole, (c) 49.4 μmole.

To calculate the electric potential, we'll use the Nernst equation
(Equation 23.13):

$$E = -2.303 \frac{RT}{nF} \log \frac{c_2}{c_1}$$

R is the universal gas constant, T is the temperature in Kelvins, n is
the number of electrons being moved as part of the redox reaction, c_1
is the concentration of the oxidized species, and c_2 is the concentration
of the reduced species.

$T = 298$ K, $R = 8.314$ J K^{-1} mole^{-1}, $F = 96,485$ C mole^{-1}, $n =$
1, $c1 = $ [NAD$^+$], $c2 = $ [NADH].

Calculating each of the above concentration: (a) -142.2 mV, (b)
63.0 mV, (c) 19.2 mV.

A.27.6 (a) Hyperfine spliting is caused by the presence of a nearby magnetic field.
Free radicals such as O$_2$$^{\bullet-}$ are paramagnetic and can cause this effect.

A.27.7 (a) The spectral width of an EPR signal depends on the freeedom of the
spin label to move in its environment. Labels added to membranes at low
temperatures have broad signals associated with rigid membranes and low
membrane fluidity. As these membranes are warmed to higher temperatures the membranes become more fluid and the spectral width of the spin
label narrows reflecting the greater freedom of motion of the label.

Exposure of membranes to superoxide anion is associate with peroxidation of the membrane, especially at unsaturated sites in the hydrocarbon
chain. The result noted in the problem suggests that the peroxidation of
the membrane leads to a loss of membrane fluidity.

(b) A thermal method such as differential scanning calorimetry could
be used to measure the membrane fluidity. Analytical techniques could be
used to show the oxidized lipid components of the membranes.

A.27.8 (a) Chemical shifts measure the difference in nuclear frequencies measured by NMR due to differences in the local electronic environment.
Electrons placed in a magnetic field will generate a small opposing magnetic field that will shield the nucleus leading to changes in the observed
nuclear frequency. (b) Ethylene (CH$_2$=CH$_2$) has the greater chemical shift
due to its double bond, which deshields the protons more than the oxygen
in dimethyl ether (CH$_3$COCH$_3$).

A.27.9 (a) J-coupling is the frequency shift due to spin-spin interactions between
nuclei. (b) (ClCH$_2$CH$_3$) Carbon 1 has two protons and carbon 2 has three
protons. The protons on carbon 1 will be split by the protons on carbon
2 and vice versa, with each group being split into $(n+1)$ peaks. Carbon
1 will be split into 4 peaks and carbon 2 will be split into 3 peaks. The
protons will not have the same chemical shift because they are in different
electronic environments. Carbon 1 has a chlorine that carbon 2 lacks.

A.27.10 (a) T_1 is the spin-lattice relaxation time. It is measured by first saturating a sample with radio energy and then removing the radiation and observing the return of the saturated frequency. The peak area measured against time will follow the exponential kinetics of a first order process and yield T_1. (b) Crystals typically have much longer spin-lattice relaxation times because their molecules have very restricted motion, which limits their ability to exchange energy with their environment (i.e. the lattice).

A.27.11 (a) Ferromagnetism, paramagnetism, and diamagnetism. Iron is ferromagnetic because it is magnetic without an external magnetic field. Paramagnetism enhances a magnetic field as it passes through a material while diamagnetism inhibits the magnetic field. (b) The magnetic field cannot last forever because everything approaches equilibrium eventually. The magnetic field will slowly lose energy through its interactions with the outside world, as well as random fluctuations within the superconducting material.

A.27.12 (a) 0.83, (b) 0.79, (c) 1.16, (d) 0.09; The answer for (c) does not make any physical sense; a system could not release more photons than it absorbs.

Chapter 28
Determining Structure by Molecular Interactions with Photons: Scattering Phenomena

28.1 Questions

Q.28.1 Show that Equation (29.12) is a general form of the Bragg equation as derived from the von Laue equations.

Q.28.2 Use the equipartition theorem to predict the energy of a standing wave in a cavity filled with standing waves of frequency, v, in thermal equilibrium.

Q.28.3 The small angle approximation states that for small angles of incidence, the sine of θ can be replaced by θ itself. Calculate and define for which small angles this statement is a reasonable simplification.

Q.28.4 What is the apparent depth of a goldfish who is 10 cm below the surface in an aquarium? (the refractive index of water is 1.33)

Q.28.5 Show that $\sin(x + vt)$ and $\sin(x - vt)$ are solutions of the wave equation.

Q.28.6 Show that the intensity of a cone of wave energy falls with a $\frac{1}{r^2}$ relationship.

Q.28.7 Calculate the length that will result in the destructive interference for a wave of frequency v.

Q.28.8 Draw a 1D slice of the amplitude and intensity of an interference pattern generated by two point x-ray sources ($\lambda = 0.1$ nm) that are 0.4 nm apart (d) at a distance of 10 mm (D) from the sources. Assume the maximum possible amplitude is 1 (in arbitrary units).

Q.28.9 A crystal is rotated from $0-30°$ in an x-ray diffractometer. ($\lambda = 1.54$ Å) How many dots of intensity will originate from the following plane spacings within first $30°$? (a) 1 Å; (b) 1.6 Å; (c) 3.3 Å.

Q.28.10 An experienced x-ray crystallographer hands you a set of disorganized data collected from an unknown 15 kDa protein crystal on an x-ray diffractometer ($\lambda = 1.54$ Å) and asks you to determine the axis lengths (a, b, c) based on the information he provided. $\theta = 8.90°$ for (5, 5, 5); $\theta = 17.06°$ for (12, 8, 9); $\theta = 12.35°$ for (3, 4, 8); $\theta = 18.45°$ for (10, 5, 11);

This chapter from *The Physical Basis of Biochemistry: Solutions Manual to the Second Edition* corresponds to Chapter 29 from *The Physical Basis of Biochemistry, Second Edition*

$\theta = 21.06°$ for (12, 20, 9); $\theta = 20.89°$ for (9, 4, 13); $\theta = 20.16°$ for (5, 5, 13); $\theta = 20.38°$ for (15, 5, 11);

Q.28.11 Draw a plane described by each of the following Miller Indices: (a) (1, −2, 0), (b) (3, 4, 0), (c) (1, 1, 1).

Q.28.13 A certain orthorhombic protein crystal has a unit cell with the following axis lengths: (3.0 nm, 5.2 nm, 4.3 nm). Find the interplanar spacing (d) for the following planes (a) (1, 0, 0), (b) (8, −4, 7), (c) (3, 6, −9), d) (3, 0, 4). Which plane do you think will have the lowest diffraction angle? Justify your answer.

Q.28.12 A researcher used Rayleigh scattering to monitor the addition of the peptide melittin, an important component of bee venom, to a mixture of 100 nm liposomes (spherical phospholipid bilayers). She found that the relative intensity of the scattered wave to the incident wave increases with increasing melittin concentration, while keeping all other experimental parameters constant; what does this suggest about the interactions between melittin and the liposomes?

Q.28.13 During a Raman scattering experiment, an incident photon has a wavelength of 10 μm and the scattered photon is 10.1 μm. Did the molecule it scattered gain or lose energy? Justify your answer.

Q.28.14 An unknown dimeric protein, isolated from an oceanic crustacean, is known to have maximum activity at 15°C, but it deactivates at 0°C and 45°C. A researcher investigates possible structural changes with respect to temperature using circular dichroism. He discovers that at 45°C, there is a substantial loss of secondary structure that is irreversible upon cooling; however, at 0°C, the secondary structure is almost unchanged and the protein's activity returns when the temperature is increased above 0°C. Propose an explanation for these observations.

28.2 Thought Exercise

It is popular in science fiction to talk about the unpredictable results that occur at a singularity where force field lines cross. Story protagonists are often found "crossing" their energy-field beam devices to invoke just such singularities (in "Ghostbusters" for example). Does the proscription against crossed field lines have any true natural world meaning, or is it just a formal syntactical rule for mapping a natural system into abstract state space?

28.3 Answers

A.28.1 The solution starts with the recognition that the cosines given in the von Laue equations represent coordinate transforms with respect to the axes of the crystal. These transforms are called *direction cosines* and relate the incident and diffracted beams. We write:

$$\cos^2 \alpha + \cos^2 \beta + \cos^2 \gamma = 1$$

$$\cos^2 \alpha_o + \cos^2 \beta_o \cos^2 \gamma_o = 1$$

If one line makes an angle 2θ with respect to the projection of the other then :

$$\cos = 2\theta = \cos \alpha \cos \alpha_o + \cos \beta \cos \beta_o + \cos \gamma \cos \gamma_o$$

Now we square the von Laue equations yielding:

$$\frac{l^2\lambda^2}{c^2} = \cos^2 \gamma - \cos \gamma \cos \gamma_o + \cos^2 \gamma_o$$

$$\frac{h^2\lambda^2}{a^2} = \cos^2 \alpha - \cos \alpha \cos \alpha_o + \cos^2 \alpha$$

$$\frac{k^2\lambda^2}{b^2} = \cos^2 \beta - \cos \beta \cos \beta_o + \cos^2 \beta$$

Summing these three equations and using the direction cosine identities gives,

$$(\frac{m^2}{a^2} + \frac{n^2}{b^2} + \frac{d^2}{c^2})\lambda^2 = 1 - 2(\cos \alpha \cos \alpha_o + \cos \beta \cos \beta_o + \cos \gamma \cos \gamma_o)$$

$$= 2(1 - \cos 2\theta).$$

Using the identity that $1 - \cos 2\theta = 2 \sin^2 \theta$ we can write:

$$(\frac{h^2}{a^2} + \frac{k^2}{b^2} + \frac{l^2}{c^2})\lambda^2 = 4 \sin^2 \theta$$

which is the Bragg equation as sought:

$$\lambda(\frac{h^2}{a^2} + \frac{k^2}{b^2} + \frac{l^2}{c^2})^{1/2} = 2 \sin \theta$$

A.28.2 The expression for the energy of a standing waves is:

$$\bar{E} \propto E_{02} + B_{02}, \text{ therefore } \bar{E} = kT.$$

A.28.3 Sine functions can be expanded using a Taylors series: $\sin(x) = (x - x^2/3! + ...)$ For very small values of x, $x^3/3!$ is small enough to be neglected. Whether the proper application of this approximation fails is determined by the researcher because it is the researcher who must set a threshold of acceptable deviation from the actual values. We'll set it at 1%. As long as $x^2/3! < 0.01 \times x$, we stipulate that x is an acceptable approximation of $\sin(x)$. To solve for the value of x where this is no longer true, we must solve:

$$x^2/3! = 0.01 \times x$$
$$x^3/6 = 0.01 \times x$$
$$x^2 = 0.06 \times x$$
$$x^2 = 0.06$$
$$x = 0.24$$

As long as the absolute values of x is less than 0.24, assuming $\sin(x) = x$ is acceptable.

A.28.4   Start with $\frac{n_1}{s} + \frac{n_2}{s'} = \frac{n_1 - n_2}{r}$. Since the surface is flat, the radius of curvature is 0 and we can write: $\frac{n_1}{s} + \frac{n_2}{s'} = 0$ Thus $s' = -\frac{n_2}{n_1}s$. Substituting gives $\frac{1}{1.33}(.01 \text{ m}) = 7.5 \text{ cm}$.

A.28.5   The wave equation is

$$\frac{\partial^2 y}{\partial x^2} = \frac{1}{v^2} \frac{\partial^2 y}{\partial t^2}$$

$$d^2y/dx^2(\sin(x + vt)) = 1 \times 1 \times \sin(x + vt) = \sin(x + vt)$$

$$d^2y/dx^2(\sin(x + vt)) = v \times v \times \sin(x + vt) = v^2 \times \sin(x + vt)$$

$$\sin(x + vt) = (1/v^2)(v^2 \times \sin(x + vt))$$

The derivation for $\sin(x - vt)$ is identical.

A.28.6   Intensity (I) is the power per unit area. $I = P/A$. The area of the cone-shaped wave front will be circular and the area of a circle is πr^2. $I = P/\pi r^2$ for a cone, so the intensity will decrease with a $1/r^2$ relationship.

A.28.7 $\Delta L = (n + 1/2)\lambda$

use $\Delta L = \lambda/2 = 90° = $ destructive
$\Delta L = n\lambda = $ constructive
maximum $= d \sin \theta = n\lambda$
minumum $= d \sin \theta = (n + 1/2)\lambda$

A.28.8   To draw a 1D slice, you have to graph the following functions over reasonable values of X. Since D is 10 mm, we will draw the graph with values of x from 0–5 mm.

$$A = 2A_o \cos \frac{1}{2} \frac{\pi x d}{\lambda D}$$

$$I = 4A_o^2 \cos^2 \frac{1}{2} \frac{\pi x d}{\lambda D}$$

Because we know the lineshapes are going to be cos and $\cos^2$, you only need to calculate a few data points and then draw the rest of the graph

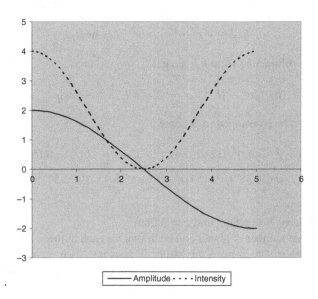

——— Amplitude · · · · Intensity

A.28.9 Bragg's law is: $2d \sin \phi = m\lambda$ Rearranging the equation we have $\varphi = \arcsin(m\lambda/2d)$. $\lambda = 1.54$ Å, m is the order of the reflection and can be any integer, although we only need to try numbers until the resulting angle is at or above 30°. Which of the following spacings will diffract in the first 30°? (a) 1 Å; (b) 1.6 Å; (c) 3.3 Å. (a) 0 dots. It's first order diffraction is at 50.35°. (b) One dot. Its first order is at 28.77°, but its second order is 74.25°. (c) 2 dots. Its first order is at 13.49°, and its second order is 27.82°. It's third order is 44.43°.

A.28.10 The von Laue equation for a orthorhombic crystal is:

$$\lambda \left(\frac{h^2}{a^2} + \frac{k^2}{b^2} + \frac{l^2}{c^2} \right)^{1/2} = 2 \sin \theta$$

We need to solve for (a, b, c); we have $\lambda = 1.54$ Å, and eight (h, k, l)-theta pairs. Because the size of the protein suggests the size of the unit cell will be in nanometers, we will convert the wavelength to nanometers have $(\lambda = 0.54$ nm$)$. We only require three sets of data to solve for three unknowns, but if we organize the data, we can see that we can simplify our calculations if we compare the correct pairs of data points.

Changes only h: $\theta = 18.43°$ for (10, 5, 11); $\theta = 20.38°$ for (15, 5, 11);
Changes only k: $\theta = 17.06°$ for (12, 8, 9); $\theta = 21.06°$ for (12, 20, 9);;
Changes only l: $\theta = 8.90°$ for (5, 5, 5); $\theta = 20.16°$ for (5, 5, 13);;

Considering these pairs of data, we are solving two equations for each unknown instead of solving three equations for three unknowns simultaneously. Rearranging the von Laue equation, we have

$$h^2/a^2 + k^2/b^2 + l^2/c^2 = (x \sin \theta/\lambda^2)$$

If we subtract $h1$ from $h2$, we get

$$(h_2)^2/a^2 - (h_1)^2/a^2 = (2 \sin \theta_2/\lambda)^2 - (2 \sin \theta_1/\lambda)^2$$

which can be rearranged to become

$$a^2 = ((h_2)^2 - (h_1)^2/((2 \sin \theta_2/\lambda)^2 - (2 \sin \theta_1/\lambda)^2) = ((15)^2 - (10)^2)/$$
$$((2 \sin(20.38°)/(0.154))^2 - (2 \sin(18.43°)/(0.154)^2) = 34.75$$

$a = 5.9$nm.
Similarly, $b = 6.8$nm and $c = 3.0$nm.

A.28.11 An infinite number of planes can be drawn for each of these Miller Indices. The simplest is to convert the Miller Indices into their reciprocals: (a) (1, $-1/2$, ?), (b) (1/3, 1/4, ?), (c) (1, 1, 1). The reciprocals can be used as the intercepts of the planes on the axes. The following graph shows (a) and (b):

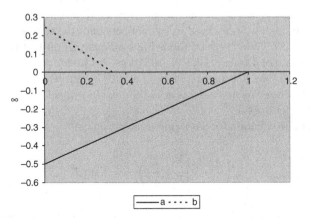

(c) is a three-dimensional plane:

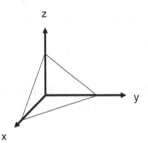

A.28.12 Use $\frac{1}{d^2} = \left(\frac{h^2}{a^2} + \frac{k^2}{b^2} + \frac{l^2}{c^2}\right)$ (a) $d = 1.73$ nm, (b) $d = 0.56$ nm, (c) $d = 0.62$ nm, (d) $d = 0.86$ nm. The diffracted angle is inversely proportional to d, so (a) will have the lowest diffracted angle.

A.28.13 Rayleigh scattering is inversely proportional to the radius of scattering particles. The increasing relative intensity suggests that the particles are decreasing in radius with increasing melittin concentration.

A.28.14 The energy of a photon is directly proportional to frequency and inversely proportional to wavelength. Since the wavelength of the scattered photon is higher than the incident photon, the scattered photon has lower energy. This energy was absorbed by "the molecule" so it gained energy.

A.28.15 At 45°C, the protein's permanent loss of structure and activity suggests that it irreversibly denatures. The retention of the protein's structure at 0°C suggests the deactivation is due to something other than denaturation. Because it is dimeric, the dimer may dissociate at lower temperatures in a way that retain the secondary structure, allowing it to reform when the temperature is increased.

Chapter 29
Analysis of Structure – Microscopy

29.1 Questions

Q.29.1 We recognize that each level of exploration carries a cost vs. benefit problem. Perhaps this is best called the "forest or trees" conflict. With greater visualization of detail there is a loss of knowledge of the overall organization of the system. This is a common problem in microscopy. What are the problems to be considered in histochemical analysis or the use of fluorescent probes. What is gained and what is lost with each of these observational levels. Is one level better than any other?

Q.29.2 A cell of 25 microns is focused by a thin lens onto the image plane. The distance between the cell and the focal point of the lens is 5 mm. The focal point of the lens is 5 cm.

   (a) What is the magnification of the lens?
   (b) What is the size of the cell image in the image plane of the lens?
   (c) Is the image right-side up or down?

Q.29.3 If the lens in Q.29.2 were to be used in a compound microscope as the objective lens along with an eyepiece lens that has a magnification of 10×, what will the size of the image formed on the retina be? What is the magnification of this instrument?

Q.29.4 Explain why monochromatic light is preferred in research microscopy for examining objects on the cellular dimension.

Q.29.5 Compared to the focal point of a beam of laser light formed by a convex thin lens at 760 nm, will the focal point of a 460 nm laser be different? Where will the focal point(s) of the two lasers be relative to one another? What is the relevance of your answer to microscopy?

Q.29.6 Resolution is often practically calculated by the formula $r = \frac{0.61\lambda}{NA}$. Why does increasing NA lead to improved resolution?

This chapter from *The Physical Basis of Biochemistry: Solutions Manual to the Second Edition* corresponds to Chapter 30 from *The Physical Basis of Biochemistry, Second Edition*

P.R. Bergethon, K. Hallock, *The Physical Basis of Biochemistry*,
DOI 10.1007/978-1-4419-7364-1_29, © Springer Science+Business Media, LLC 2011

Q.29.7 What is the effect on the resolution of an optical microscope if monochromatic light of 700 nm is used for illumination? 650 nm? 550 nm? 450 nm? 360 nm?

Q.29.8 What histochemical stains would be useful for identifying the following substances in a biological specimen? (a) lipid containing organelles; (b) amyloid fibrils; (c) acidic mucopolysaccharides; (d) nucleic acids; (e) collagen fibers.

Q.29.9 You are asked to experimentally evaluate a new multi-stage microfluidic fuel cell in which a crude solution of cane sugar (sucrose) is first hydrolyzed by and acidic stage and then glucose and fructose diffuse into a second stage where electrochemical oxidation produces power. Propose a microscopic method to measure the rate of sucrose hydrolysis in the first stage.

Q.29.10 What are the advantages of a ratiometic fluorescent dye over a non-ratiometric probe?

Q.29.11 Both Fura and Indo are fluorometric probes that can be used to image intracellular calcium concentration. Can the same microscope optical train be used for both of these probes?

Q.29.12 A scanning tunneling microscope is set to operate in constant height mode at 0.9 nm above the surface with a tip bias of 0.01 V. A scan is performed in the +x direction. What topography do the following measured currents (in picoamperes) suggest?

Distance along x axis (nanometers)	Tunneling current (picoamperes)
0.0	100
0.05	1000
0.10	86,000
0.15	100,000
0.20	86,000
0.25	1,000
0.30	100

29.2 Answers

A.29.1 The tradeoffs that come with:

   Histochemical analysis stem from the use of chemical reactions with a specimen that must be prepared so that it can be a proper environment for the requisite chemistry to occur. Important information related to chemical composition and its spatial arrangement is gained in histochemical analysis and the problem of increasing contrast which helps solve the visualization problem in general is solved. However the added information comes at the cost of "fixing" or preparing a sample that may well change its spatial relationships thus causing visual artifacts that may be attributed to real structural information when they are in fact noise. The chemical reactions will have a variability that may influence what can be learned about the concentrations of "imaged molecules" as well as their locations. Diffusion and photobleaching of the reactants (which are the observables in the system) will induce noise in the measurement of both location and amount of substrate.

Fluorescent probes have many of the same advantages and disadvantages of histochemical analysis however because they can be used in systems in which a condition of very low background light can be established, a very high signal to noise ratio can be obtained. This can make them very sensitive as probes. Fluorescent probes are often used as functional reporters sensing chemical conditions (pH, membrane voltage, ion concentrations, enzyme activities) and physical environments (membrane anisotropy, fluidity, diffusion constants). However in order to report on these physiochemical properties of systems under investigation, they must interact with the environment and may well change the environment itself and suffer from an often unknowable degree of "observer-system" coupling. For example most pH probes are themselves buffers and can alter the pH of the probed environment, membrane probes are cation/anions that may induce Donnan effects, ion probes measure through chelation of the ions of interest, enzyme probes are substrates and may be inhibit, compete or have allosteric effects on the enzyme being probes and in the case of membrane probes, themselves change the behavior of the very membrane they are reporting on because of incorporation into the membrane under study.

Whether one level of investigation is better than another is dependent on the question that is being asked and the requirements of the hypothesis that is being tested or information that is needed to be obtained. It is always important to include the overall costs of experimenter time (preparation), experimental cost (equipment and reagents), the time required for an observable to be measured (where photobleaching or sample destruction may become important) and whether the analysis requires destruction of the system under observation. The goodness of the level of observation is dependent on the questions posed in Chapter 3 about the goodness of the model.

A.29.2 (a) To answer this question use the magnification formula $M = -\frac{f}{a}$ where f is the object plane focal point and a is the distance from the object to the focal point. $M = -\frac{f}{a} = -\frac{50 \times 10^{-6} m}{5 \times 10^{-6} m} = 10$

 (b) The size of the image is $10 \times 25\,\mu m$ or $250\,\mu m$.

 (c) The image is upside down as indicated by the (−) sign in the formula.

A.29.3 The magnification of the microscope is the objective power (10×) multiplied by the eyepiece magnification (10×) or 100× magnification. The size of the final image is $250\,\mu m \times 10$ or 2.5 mm.

A.29.4 Because the degree of refraction of light through a lens is wavelength dependent, if polychromatic is used to illuminate a specimen, there will be some chromatic aberration in the final image even with the best achromatic lens systems. When monochromatic light is used, there are no other wavelengths of light that will distort the image due to the wavelength dependent refraction.

A.29.5 The two focal points will not be in the same place. The 460 nm laser will have a focal point closer to the lens than the 760 nm laser beam. The wavelength dependent refraction by the lens is the cause of chromatic aberration.

A.29.6 The numerical aperture (NA) is a measure of the number of orders of diffraction that can be captured by a lens, $NA = n \sin u$. Since the image information is carried in all of these orders of diffraction, the wider the collection angle u, the greater the magnitude of NA and the higher the resolving power, i.e. the smaller the distance between two objects' set of Airy discs needed to visually separate them.

A.29.7 To see the effect of changing the wavelength light on resolution, use $r = \frac{0.61\lambda}{NA}$ and a fixed NA (0.95). This gives the following result: 700 nm is used for illumination? 650 nm? 550 nm? 450 nm? 360 nm?

(a) $r_{700} = \frac{0.61(700 \times 10^{-9}\text{m})}{0.95} = 0.45 \times 10^{-6}\text{m}$;

(b) $r_{650} = 0.42 \times 10^{-6}\text{m}$; (c)$r_{550} = 0.35 \times 10^{-6}\text{m}$; (d)$r_{450} = 0.29 \times 10^{-6}\text{m}$; (e)$r_{360} = 0.23 \times 10^{-6}\text{m}$

A.29.8 (a) sudan black or oil red O are stains that react with neutral lipids and triglycerides in cells; (b) congo red binds to amyloid fibrils and then can be used diagnostically by showing an apple-green birefringence under conditions of polarizing microscopy; (c) methylene blue stains acidic mucopolysaccharides; (d) heamtoxylin will stain nucleic acids; (e) eosin reacts with and stains collagen fibers.

A.29.9 A polarization microscope can be used to evaluate the hydrolysis that should be occurring in the first stage of this microfluidic device. The intensity of transmitted light through these microfluidic stages can be measured with a CCD, digitizing camera or appropriate photomulitplier system. The polarizer at the condenser level could be set to transmit polarized light at +66.5° and the analyzer can be set to transmit this plane of light. As the hydrolysis proceeds, the light will be levorotated and the intensity of the light will decrease in proportion. The reaction could be followed by an increase in light intensity if opposite polarization parameters are chosen.

A.29.10 A ratiometic dye while more complex to use in terms of equipment set up and computation of data points has great advantages in the quality of the data that results. Ratioing internally controls for the artifactual effects that will occur with uneven dye loading and uneven cell geometry, especially thickness. Ratioing also reduces the signal noise caused by leakage of dye and photobleaching.

A.29.11 While both of these ratiometric probes are can be used in a fluorescent microscope, ideally with an epifluorescent train, Indo probes require a single excitation wavelength to be provided and the change in intensity of two emission wavelengths to be measured while Fura probes require two excitation wavelengths while measuring the change in intensity of a single wavelength. Thus the epifluorescent microscope optical train used for Indo requires only a single laser or excitation filter/lamp setup but must either have two separate emission channels (two filtered paths) or switch filters in a single path. Fura requires two excitation sources (two lasers or a switched excitation filter setup) and a single emission-photomultiplier path for measurement.

A.29.12 The variation in the current recorded follows a sinusoidal pattern suggesting that the surface topography is regularly varying in a hills and valley type of structure with the x dimension periodicity on the order of 0.3 nm and vertical rise on the order of 0.5 nm.

Distance along x axis (nanometers)	Tunneling current (picoamperes)	Tunneling gap (nanometers)	Rise of surface (from tip in nm)
0.0	100	0.7	0
0.05	1000	0.45	0.25
0.10	86,000	0.27	0.43
0.15	100,000	0.2	0.5
0.20	86,000	0.27	0.43
0.25	1,000	0.45	0.25
0.30	100	0.7	0

The data provided is only for a single scan in the x-direction so it is not possible to determine from this data whether the variation suggests regular conical variation (like an egg crate pattern) or some other periodic pattern. In addition because the current falls exponentially with increasing distance between the STM tip and the conductive surface, the constant height mode is less sensitive to detecting variations in surface topography as the surface falls away from the tip.

Part VI
Physical Constants

Chapter 30
Physical Constants

Atomic mass unit	u	1.661×10^{-27} kg
Avogadro's number	N_A	6.022×10^{23} mol^{-1}
Bohr magneton	μ_B	9.27×10^{-24} JK^{-1}
Boltzmann constant	k	1.381×10^{-23} JK^{-1}
Electron rest mass	m_e	9.00×10^{-31} kg
Elementary charge	e	1.602×10^{-19} C
Faraday constant	F	9.6485×10^4 C mol^{-1}
		2.306×10^4 cal mol^{-1} eV^{-1}
Neutron mass	m_n	1.673×10^{-27} kg
Permeability of free space	μ_0	$4\pi \times 10^{-7}$ T - m A^{-1}
Permittivity of vacuum	ϵ_0	8.854×10^{-12} C^{2}N^{-1}m^{-2}
Physical Constants		
Planck constant	h	6.626×10^{-34} J s
Proton rest mass	m_p	1.673×10^{-27} kg
Speed of Light (in vacuum)	c	2.990×10^8 m s^{-1}
Universal gas constant	R	8.314 J K^{-1} mol^{-1}
		1.987 cal K^{-1} mol^{-1}
		0.082 L atm K^{-1} mol^{-1}

30.1 Conversions

1 Joule	$= 1$ Newton meter
1 atmosphere	$= 1.01325 \times 10^5$ Pa (Pascal)
1 liter	$= 1 \times 10^{-3}$ m^3

P.R. Bergethon, K. Hallock, *The Physical Basis of Biochemistry*,
DOI 10.1007/978-1-4419-7364-1_30, © Springer Science+Business Media, LLC 2011

Index

A

Ab initio method, assumptions/constraints, 75
Activated complex, 101
Activation energies, 97–98
All atom force field, 75
Alzheimer's disease, 73
Amyloid plaques, primary component of, 73
Approximate laws, reason for using, 11
Aqueous solution, lipids in, 71–72
Archimedes' principle, 111
Argon, 33
Arrhenius equation, 99

B

Balmer series wavelength prediction, 36
Bathochromic shift, 116
Battery, work performance of, 27
    power delivered by, 27
Beer-Lambert law, 117
Benzene, 48
Binomial distribution, 15
Biochemist knowledge, 12
Bioelectrochemistry, 103–105
Biological systems
    charge transfer in, 103–105
    under study, overview of, 9
Biomolecules, separation/characterization,
        109–114
Biophysical chemistry, 12
Biophysical forces in molecular systems,
        29–33
Blood
    flow, processes of intellectual computation,
        13
    ionic strength of serum component
        of, 67
    pH, zeta potential, 104
Boltzmann constant, 40
Bond energies, bond type, 44–45

Born approximation, 66
Born model, assumptions, 65
Born-Oppenheimer approximation, importance
        of, 39, 41
Bragg equation, 121
Bragg's law, 125
Brain system/boundary/surroundings, 44
Breathing modeled using ideal Pressure-
        Volume work, 25

C

Carbonic anhydrase
    enzyme efficiency, 97
    limiting factor, 99
Carbon monoxide, 54
Catalase
    enzyme efficiency, 97
    limiting factor, 99
Cell
    membrane, 9
    subsystems of
        cytosol, 9
        mitochondrion, 9
        ribosome, 9
    system/boundary/surroundings, 45
Cellular membrane channel, binomial
        distribution, 16, 18
Charge-separation reaction, 105
Chemical kinetics, 97–101
Chemical potential field, flow in, 89–90
Chemical principles, 39–41
Cholesterol, 72
Chymotrypsin
    enzyme efficiency, 97
    limiting factor, 97
Circular dichroism, 113
Classical mechanics' fundamental view of
        state space, 35
Clausius-Clapeyron equation, 58